Stone decay in the architectural environment

Edited by

Alice V. Turkington
Department of Geography
University of Kentucky
1473 Patterson Office Tower
Lexington, Kentucky 40506-0027
USA

THE
GEOLOGICAL
SOCIETY
OF AMERICA

Special Paper 390

3300 Penrose Place, P.O. Box 9140 ▪ Boulder, Colorado 80301-9140 USA

2005

Copyright © 2005, The Geological Society of America, Inc. (GSA). All rights reserved. GSA grants permission to individual scientists to make unlimited photocopies of one or more items from this volume for noncommercial purposes advancing science or education, including classroom use. For permission to make photocopies of any item in this volume for other noncommercial, nonprofit purposes, contact the Geological Society of America. Written permission is required from GSA for all other forms of capture or reproduction of any item in the volume including, but not limited to, all types of electronic or digital scanning or other digital or manual transformation of articles or any portion thereof, such as abstracts, into computer-readable and/or transmittable form for personal or corporate use, either noncommercial or commercial, for-profit or otherwise. Send permission requests to GSA Copyright Permissions, 3300 Penrose Place, P.O. Box 9140, Boulder, Colorado 80301-9140, USA.

Copyright is not claimed on any material prepared wholly by government employees within the scope of their employment.

Published by The Geological Society of America, Inc.
3300 Penrose Place, P.O. Box 9140, Boulder, Colorado 80301-9140, USA
www.geosociety.org

Printed in U.S.A.

GSA Books Science Editor: Abhijit Basu

Library of Congress Cataloging-in-Publication Data

Stone decay in the architectural environment / edited by Alice V. Turkington.
    p. cm. — (Special paper / Geological Society of America ; 390)
  Includes bibliographic references.
  ISBN 0-8137-2390-6 (pbk.)
    1. Building stones--Deterioration. 2. Stone buildings--Conservation and restoration. I. Turkington, Alice V. II. Special papers (Geological Society of America) ; 390.

TA426.S764 2005
691'.2--dc22

                                         2005047700

Cover: Top right: Weathered eleventh century carving on Hayles Abbey, Cotswolds, UK. Bottom left: Ruins of Karnack Temple near Luxor, Egypt. Bottom right: Black crust development on a limestone balustrade, Oxford, UK.

Contents

Preface: Introduction to stone decay in the architectural environment

Stone is not only the material of many significant buildings; it comprises many historic, cultural, and artistic artifacts. Much of our knowledge of civilization, from ancient through historic times, is based on artifacts that have been written on, sculpted from, or built from stone. Anthropogenic impacts have accelerated the decay of many stone structures, for example through high urban pollution levels and consequent acid deposition. Conversely, interest in the past has blossomed, and the past may be viewed as a "foreign country" with its own booming tourist industry. The material links to our past are often comprised of stone, and conservation or preservation of these structures has gained increased social, cultural, and financial importance. Given the uniqueness of many monuments, combined with their universal appeal, stone conservation may be viewed as a truly global concern.

Conservation of stone structures requires an understanding of the physical, chemical, and biological processes causing decay and the effects of restoration strategies. Groups conducting research on stone decay and conservation include geologists, geomorphologists, conservation architects, and engineers; cross-fertilization of ideas between the interested groups is increasing, yet there is an urgent need for increased collaboration and multidisciplinary research. There is a fundamental need to integrate our knowledge of the entire decay system, including the stone properties, environment, decay processes, forms produced, and preservation goals and strategies, to develop a new theoretical framework for stone decay and conservation research.

The work that contributes to this volume samples state-of-the-art research on urban stone decay, and research on stone decay and conservation has never been more active. A huge volume of literature pertaining to pollution-related damage to cultural heritage has been generated in the past decade; much of it is sourced in proceedings of international conferences and internationally funded projects. The papers in this volume are presented by geologists and geomorphologists, whose various approaches to diagnosing and predicting stone decay are pushing the boundaries of their traditional disciplines.

Numerous lessons on rock weathering can be learned from investigating stone decay in the architectural environment, not least because of a relatively sound knowledge of dates of exposure and changing atmospheric conditions. As Smith et al. (Chapter 1) suggest, strategic sampling from buildings, complemented by standard petrographic, chemical, and visual analyses can elucidate many of the complexities of the weathering system. The weathering system is not as simple as it may appear, however. Many observations of weathering processes and rates have demonstrated the nonlinearity of stone response to environmental stresses, and of stone response to a combination of processes acting in concert in natural or urban environments. Few researchers have analyzed this problem from a nonlinear dynamical systems perspective, yet it is of paramount importance if prediction of stone decay is to improve. Viles (Chapter 2) demonstrates that stone response to decay processes is inherently nonlinear and often chaotic, a theoretical framework that has profound implications for both predicting and interpreting stone decay.

The destructive effect of air pollution on our stone heritage has long been apparent and inspired widespread interest in the past 30 years as the destructive effects of acid rain became evident. While accelerated dissolution of limestone structures (e.g., Venice) and black crust growth on calcareous stone surfaces (e.g., Oxford) has received a lot of attention, not all stones have responded to acidic precipitation in a similar manner. Indeed, many stones continue to decay at a high rate due to the "memory effect" of past aggressive atmospheric conditions. Meierding (Chapter 3) presents a geographic survey of serpentine stone buildings affected by acidic precipitation in Pennsylvania, providing a database on which future studies can build. Meierding used the state of decay of marble tombstones as a surrogate measure of pollution concentrations, a method he pioneered. The assessment of tombstones as indicators of weathering rates in a range of environments has often been documented, but the methodologies used have been based on a number of assumptions, many of which may not be valid. Roberts (Chapter 4) explicitly addresses these issues and provides an informative critique of the methodological opportunities and pitfalls these assumptions generate.

Decay of one of the world's most spectacular stone structures, the city of Petra, has been the focus of 13 years of research by Paradise (Chapter 5), who synthesizes his research to date to draw out the main themes and questions raised by this research. The sandstone façades of the city have been decaying at an accelerated rate, largely due to visitor pressure; Paradise's work illustrates how stable this 2000-year-old piece of antiquity has been until it was recently rediscovered by the mass tourism industry. For those structures undergoing severe decay, replacement of stones is a major part of restoration efforts. The choice of stone for construction and conservation is a thorny issue, as most durability tests are semiquantitative at best and inaccurate at worst. One key characteristic of a durable building stone is its resistance to volumetric pressures exerted by salt or ice crystallization and by swelling of the clay matrix. A new method to assess these parameters is proposed here by Scherer and Jimenez Gonzalez (Chapter 6), who have reduced a highly technical analysis to a simple, replicable procedure to assess the nature and rate of swelling in clay-bearing stone.

Stone heritage is constantly under threat from natural and anthropogenic decay processes, yet stone buildings, structures, and artworks will undoubtedly remain a permanent feature of our culture. It is clear that geological approaches to investigations of stone decay will remain at the forefront of future research, as we progress in our understanding of the processes, controls, and responses within the stone decay system, and of their variations and interactions in time and space. We hope that this volume will inform earth scientists concerned with rock weathering in natural and urban locales, and that it might be of benefit to scientists, students, conservators, and practitioners whose interest lies at the interface between research and its application.

Alice V. Turkington

Geological Society of America
Special Paper 390
2005

Urban stone decay: The great weathering experiment?

Bernard J. Smith*
School of Geography, Queen's University Belfast, Belfast BT7 1NN, UK

Alice V. Turkington
Department of Geography, University of Kentucky, Patterson Office Tower, Lexington, Kentucky 40506-0027, USA

Joanne M. Curran
Stone Conservation Services Ltd., The Gas Office, 4b Cromac Street, Belfast BT7 2JD, UK

ABSTRACT

This paper outlines recent progress in modeling salt-induced decay of urban stone, using both field-based and laboratory-based approaches. The opportunities afforded by examining stone structures of known age, composition, and exposure history are highlighted, and the fruitfulness of multidisciplinary research in this area for geologists, geomorphologists, engineers, and conservation architects is demonstrated. Examples are given that show how studies of urban stone decay have informed our understanding of salt weathering, surface loss, and both meso- and micro-environmental controls on weathering. Institutional and cultural reasons for the lack of discourse between building stone researchers and those concerned with natural rock outcrops are discussed, and the benefits of information exchange between the two disciplines is advocated.

Keywords: urban stone decay, rock weathering, salt weathering, simulation experiments, damage mapping, spatial variation, exposure trials, subsurface decay, salt content, salt type, salt distribution, microclimate, environmental controls, gravestones, positive feedback, damage assessment.

INTRODUCTION

An increasing catalogue of projects and publications are available in which earth scientists have sought to apply their knowledge of natural stone weathering to the explanation of building stone decay (e.g., Smith, 2003; Cooke and Gibbs, 1993). Geologists and geomorphologists have increasingly made use of the educational and research potential of stone used in construction, accessing this resource to gain insights into the nature and controls upon the weathering and erosion of stone. There is a considerable history of the use of urban environments as well-ordered museums of different stone types, neatly polished and freshly exposed at a known date. For example, an ongoing interest in the use of gravestones to synthesize erosion rates over time goes back to the pioneering work in Scotland of Geikie (1880). As geologists and geomorphologists have adopted buildings as "natural laboratories," there are still relatively few multidisciplinary groups examining the various aspects of urban stone decay (see Siegesmund and Weiss, 2002; Smith and Turkington, 2004). This paper presents an overview of research conducted in Queen's University of Belfast in collaboration with local conservation architects, research specifically concerned with assessing

*E-mail: b.smith@qub.ac.uk

Smith, B.J., Turkington, A.V., and Curran, J.M., 2005, Urban stone decay: The great weathering experiment? *in* Turkington, A.V., ed., Stone decay in the architectural environment: Geological Society of America Special Paper 390, p. 1–9, doi: 10.1130/2005.2390(01). For permission to copy, contact editing@geosociety.org. ©2005 Geological Society of America.

stone damage, diagnosing decay processes, and understanding the complex operation of the stone decay system.

The relatively controlled conditions of stone placed in buildings has considerable potential to provide insights into how rocks behave in natural outcrops. It may be argued that "weathering experiments" are created by architects who simultaneously expose combinations of stone types to the same environmental conditions (e.g., Cooper et al., 1991) or conveniently align structures of the same stone to allow evaluation of aspect effects. These "experiments" are further facilitated by an appreciation of the nature of atmospheric conditions in many cities during the period of exposure and by opportunistic sampling during restoration or conservation treatments. These "weathering experiments" are not, however, as simple as they first appear. For example, individual stone blocks are emplaced in mortar, which may vary widely in composition and potential deleterious effects on stone. Microclimatic conditions around a building are complex, affected by local urban geometry and building design. The stone in question also develops a unique stress history, originating from the time of construction. Warke (1996) categorized the stress histories of building stones as responses to pre- and post-emplacement factors. The pre-emplacement factors include effects resulting from quarrying, curing, and preparation procedures including the finish of the stone surface. The post-emplacement factors encompass, most importantly, legacies of previous pollution conditions including the build up of salts, most notably gypsum. This gives the stone a "memory" of past conditions and it is this that is responsible for the continuation of decay even after nineteenth and twentieth century atmospheric pollution levels have been reduced. It is also the reason why replacement stone rarely weathers to resemble original stonework that has a different stress history or "memory."

Natural rock weathering and urban stone decay have many similarities, such as stone surface microclimatic conditions, interactions between chemical, physical, and biological processes, and the weathering forms that result from material loss. Investigation of both urban and rural stone deterioration has utilized three main methodologies: field observations and sampling, exposure trials, and laboratory-based simulations of weathering. This paper will focus on research by a multidisciplinary group using each of these approaches to examine the nature of salt-induced decay of stone, particularly sandstone, in polluted urban environments.

URBAN STONE DECAY AND NATURAL ROCK WEATHERING

Winkler (1975) and Smith (2003) have both argued that building façades that efficiently shed water have many characteristics in common with natural rock outcrops. In particular, there is a marked commonality between conditions experienced on buildings and those found in hot desert environments. These similarities include:
- Large expanses of bare, unshaded stone;
- Periodic wetting;

- High absolute temperatures;
- High surface temperature ranges;
- Rapid surface drying by strong winds;
- High concentrations of atmospheric dust;
- An abundance of salt and the importance of salt weathering.

It is hardly surprising, therefore, that arid zone geomorphology is the one area of the earth sciences that has taken a serious interest in studies of building stone decay.

Rock weathering in deserts produces spectacular landscapes and a number of distinctive forms, such as alveoli and tafoni. Small-scale variation in stone response to weathering processes has been the subject of theoretical debate concerning causative processes and controls (e.g., Turkington, 1998; Turkington and Phillips, 2004). On a building, subtle variations in physical makeup, chemistry, and mineralogy exist that can be exploited both by weathering processes and by weathering researchers examining the controls exerted by these properties. A recent example of this is the study by Turkington and Smith (2004) that took advantage of the randomly selected arrangement of sandstone blocks in the wall of a church to map weathering features associated with individual stones. By assessing the degree of connectivity between these features (i.e., the number of edges adjacent to blocks exhibiting similar patterns of decay and/or alteration) it was possible to establish, for example, that material loss through scaling, flaking, and granular disaggregation tends toward isolation and hence is dependent upon the intrinsic properties of individual blocks. In contrast, a greater degree of connectivity (clustering) was observed between blocks with green (biological) and black (pollution-derived gypsum) crusting. This suggests that these crusts are more strongly controlled by environmental variations across the wall.

The importance of surface modifications of the type described above is widely recognized, not only as the aesthetic stimulus for most stone cleaning but also because of their impact on weathering characteristics. Iron mobilization, for example, can lead to surface hardening that, if breached, can lead to rapid loss of the underlying, weakened stone. Iron coatings, case hardening, and rock varnishes are commonly observed on natural outcrops and debris; it has been suggested that surface hardening in tandem with subsurface weakening represents the most characteristic mode of sandstone deterioration in a range of environments (Robinson and Williams, 1987). Stone decay research, such as the use of sequential extraction techniques to examine iron and trace metal mobility (McAlister et al., 2003), has promoted understanding of where and how elements are held in surface crusts and under what conditions they could be mobilized.

Spatially heterogeneous weathering has been of interest to geologists and geomorphologists for centuries, but investigations have not often examined the distribution of weathering forms on a local, or outcrop, scale. Regional geographic distributions of weathering products were discussed by climatic and climatogenetic geomorphologists in the twentieth century (e.g., Peltier, 1950), but more recent analyses have been resolutely process-based. One of the major contributions that students of urban stone decay have made to weathering studies in general is the development of classification systems for weathering features which, linked to knowledge of weathering processes and mechanisms, make effective diagnostic tools. The most thorough and widely referred to of these classifications is that developed at Aachen by Fitzner and his colleagues (e.g., Fitzner et al., 1997; Fitzner and Heinrichs, 2002). However, a drawback to many of these classifications is that they tend to treat and classify blocks on a one-by-one basis, which makes it difficult to appreciate the pathology of a building as a whole. This has led others to attempt an alternative holistic approach. For example, building damage has been described by adaptation of the internationally recognized TNM (tumor, nodes, metastases) staging system used for cancer diagnosis and treatment recommendation (Warke et al., 2003). An underlying imperative behind these classifications is the possibly severe penalties, professionally and financially, of making the wrong diagnosis and/or recommending an inappropriate treatment. While many academics might view these "rigors of the market place" with apprehension, there is nevertheless much to be said for a system that concentrates the mind by providing a direct link between actions and consequences.

This brief discussion has highlighted some of the more productive areas of cross-fertilization of ideas between weathering and stone decay researchers. Many others may be documented, not least the roles of moisture and biological organisms. Stone decay researchers have made significant contributions in both areas and have recorded their findings in a series of major publications. These include detailed texts on moisture deposition, absorption, and movement by Camuffo (1998) and Hall and Toft (2002) that are particularly valuable for the understanding they provide of decay in environments where moisture is available only in limited quantities. Complementing these texts are recent edited volumes on biodeterioration (e.g., Saiz-Jimenez 2003) and others that deal with it in a wider context such as that edited by Galan and Zezza (2000), supplemented by more accessible reviews (e.g., Wakefield and Jones 1998). The next sections discuss recent developments in simulation experiments, field observations, and exposure trials of salt attack on building stone.

SIMULATIONS OF URBAN STONE DECAY

Investigations of salt weathering in hot deserts have been both inspired and informed by studies of building stone decay. This is most clearly manifested in the modification of the sodium sulfate durability test for building stone to artificially replicate weathering environments found in hot deserts, most notably initially by Goudie et al. (1970) and Goudie (1974). In this way it has been possible to compare the "performance" of different stones subject to a severe weathering environment and to begin to identify those stone properties, such as microporosity (Cooke, 1979), that might influence susceptibility to decay. Similarly, by subjecting blocks of the same stone to repeated cycles of wetting and drying using saturated solutions of different salts and salts in combination and by varying the thermal regime used, it has been possible to speculate on the broad environmental factors that control salt

weathering. By charting these simulation experiments it is also possible to see how their design has continued to be influenced by observations of weathering on buildings. Thus, for example, a link can be made between observations of damage to buildings caused by salts derived from the capillary rise of groundwater (e.g., Arnold, 1982; Cooke et al., 1982; Amoroso and Fassina, 1983) and the use by Goudie (1986) of the "wick effect" to reproduce salt weathering under laboratory conditions. Conversely, in stonework and outcrops above the limits of groundwater rise, it has been observed that salts typically accumulate gradually and are repeatedly wetted by precipitation that is at best a very dilute salt solution. This has lead researchers to question the relevance of repeatedly immersing test blocks in salt solutions and to experiment with the one-off loading of blocks in a saturated salt solution and subsequent cycling with either dilute salt solutions (Smith and McGreevy, 1983) or water (Goudie, 1993). More recently, a commercial salt corrosion cabinet has been used to moisten one exposed surface of test blocks with a dilute salt solution mist that has successfully reproduced surface scaling. This is considered to replicate moistening of the stones similar to that produced by condensation, not only on sheltered areas of buildings but also in, for example, coastal deserts (Smith et al., 2002).

A further consequence of treating samples with concentrated salt solutions is that once near-surface pores are loaded with salt, subsequent moisture penetration may be restricted, and studies by Smith and Kennedy (1999) have shown that moisture uptake by experimental blocks can be modified after as few as three salt applications. This suggests that the nature of salt weathering processes can change as salt attack proceeds and that salt accumulation might inhibit infiltration, especially where moisture is only available in small quantities from, for example, dewfall. One consequence of this surface modification is that it can significantly reduce surface porosity and permeability with important implications for future patterns of moisture and salt ingress and long-term decay (Curran and Smith, 2000).

Simulation experiments tend to rely upon a simple, objective measure of durability or change (e.g., weight loss), which ignores the fact that when such durability tests were initially developed at the UK Building Research Establishment, emphasis was placed on the role of the experienced investigator in interpreting the significance of the weathering forms produced and determining whether the tests are producing "real" weathering (Inkpen et al., 2004). While intuitive tests of comparability are important, ultimately it is important that weathering responses are themselves quantified to facilitate objective assessment of simulation procedures. Indeed, through such quantification, it may eventually be possible to complement physical simulations with mathematical predictions using procedures such as finite element modeling (e.g., Weiss et al., 2002).

SURFACE AND SUBSURFACE STONE DECAY

The surface breakdown of rock outcrops and building stones in the presence of salts tends to be dominated by contour scaling, multiple flaking, granular disaggregation, or a combination of these. Such phenomena have traditionally been associated with near-surface cycles of heating and cooling and wetting and drying driven by diurnal temperature cycles and wetting through, for example, individual storms. This assumption is reinforced by field studies that are often restricted to surface or immediately subsurface samples for analysis and by simulation experiments designed specifically to replicate short-term patterns of wetting and drying. As a consequence, explanations of surface breakdown have primarily associated fracturing and disaggregation with zones of salt accumulation produced by the inward and outward migration of salt solutions into otherwise dry stone. Where drying is rapid, this is thought to produce salt accumulation at or near the wetting front and to ultimately result in contour scaling (Smith and McGreevy, 1988). Where drying is slower, salt solutions can migrate back to the surface, where evaporation and crystallization is thought to favor granular disaggregation (Smith et al., 1988).

Observations of actual weathering forms and patterns at various stages of development raise the suspicion that other controls may also operate. For example, on buildings across a wide range of climatic regimes, and even in hot deserts (Smith, 1994), it is noticeable that while scaling, flaking, and disaggregation can be found on any surface, they often concentrate in moist, sheltered locations where diurnal heating is ameliorated and stone remains moist longer. Added to this, it has been noted that, particularly on sandstone buildings (see Smith et al., 1994a), the delamination of individual building blocks by contour scaling can be followed by rapid retreat and complete destruction through a combination of multiple flaking and granular disaggregation. This is despite the fact that the retreating block is progressively sheltered from rain and direct insolation and that any near-surface accumulation of salt should be lost with the initial scale.

To investigate why retreat persists, complete blocks of sandstone were removed from a church undergoing restoration in the polluted maritime environment of Belfast, Northern Ireland. Chemical analysis showed them to contain a mixture of salts including calcium sulfate derived primarily from atmospheric pollution and sodium and magnesium chlorides primarily from marine aerosols. What is interesting is that sampling of entire blocks showed that the less soluble gypsum does indeed concentrate near the exposed surface where it is associated with surface-parallel cracking and scaling (Warke and Smith, 2000). In contrast, the more soluble chloride salts were found throughout the blocks to depths in excess of 20 cm (Turkington and Smith, 2000) and act as salt reservoirs that can be exploited as the blocks retreat. These relationships are illustrated in Figure 1, which shows chloride and sulfate concentrations in sandstone blocks from the church experiencing different patterns of decay. This highlights the link between chloride salts and multiple flaking characteristic of retreating blocks and the association of a mixture of salts with initial delamination. The practical importance of these "deep salts" was shown by a test wall (Fig. 2) that was "dressed back" by chiseling away surface scales and loose debris.

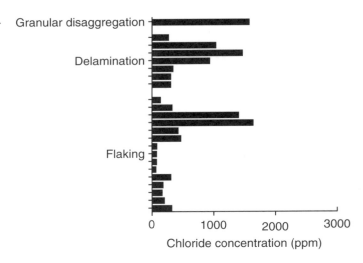

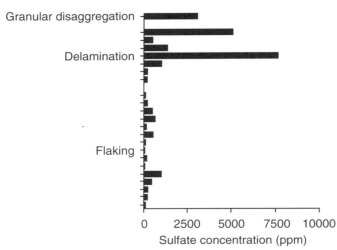

Figure 1. Chloride and sulfate ion concentrations on the surfaces of representative sandstone blocks showing granular disaggregation, flaking, and delamination (scaling), St Matthew's Church, Belfast, Northern Ireland.

Within months, a salt efflorescence formed, and active multiple flakes appeared as salts were drawn to the new surface. These observations were only possible because of the ability to remove complete blocks of stone, and to an extent it was the commercial pressures of having to identify appropriate conservation procedures that drove the direction of the work. As a consequence, it was possible to demonstrate that although the immediate cause of breakdown can be linked to near surface stressing, the underlying pattern of salt ingress, storage, and subsequent mobilization is more complex.

Chemical mapping of salt content also revealed inconsistent anion/cation ratios (Turkington and Smith, 2000), which suggests that "salts" do not necessarily penetrate through the stone solely in solution. A possibility is that "salts" are also dispersed by ion diffusion through blocks that are completely saturated

Figure 2. Trial sandstone wall at St Matthew's Church, Belfast, Northern Ireland, showing renewed flaking and salt efflorescence within four months of the original weathered surface being removed. (A) The trial wall "dressed" back to remove active scaling and flaking. (B) Detail of the same wall after four months, showing renewed decay as salts stored within the stone are brought to the surface by repeated wetting and drying.

over, in this case, the wet winter months. Such a process is fully understood by, for example, those studying chloride corrosion of concrete where the process can be measured using diffusion cells (Shaát et al., 1994) in which reservoirs of salt solution and deionized water are separated by the concrete samples to be tested. In low porosity concrete, diffusion rates are invariably very slow; however, diffusion cell measurements using 20-mm-thick discs of Permian quartz arenite (Dumfries sandstone) suggest that diffusion can be rapid through porous building stone. Tests (Smith et al., 2001) using 0.55 molar solutions of sodium chloride and sodium sulfate produced a flow rate of 93.7 ppm/day for chloride from the salt solution to the deionized water, reaching a steady-state after <24 h. Interestingly, in light of the patterns of salt distribution in the blocks taken from the church, the flow rate for sulfate was markedly lower at 16.2 ppm/day, and a steady-state was reached after ~100 h (Fig. 3). These results appear to confirm the possibility of ion diffusion within stonework and emphasize the need to think beyond near-surface mechanisms to consider, for example, seasonal climatic cycles, the effects of which penetrate deep into stone. In addition to acting as a salt reservoir, the anions mobilized by these cycles may play other roles, such as selective corrosion of quartz cements that increase susceptibility to more traditional salt weathering mechanisms in certain sandstones. Such possibilities remain largely to be tested.

ENVIRONMENT CONTROLS ON URBAN STONE DECAY

There is an unfortunate tendency in weathering studies to compartmentalize processes on the basis of macro-climatic conditions, hence the common usage of terms such as desert, periglacial, and tropical weathering in geomorphological texts. In part this is residual from a climatic approach to geomorphology that has been largely superseded—at least outside of continental Europe—by a more reductionist, process-based approach that favors a forensic examination of the processes responsible for change within landscapes. Questioning of the broad-based and largely descriptive approach taken by climatic geomorphology can be traced back to the 1960s, when reviews such as those by Common (1966) and Stoddart (1969) were especially critical of the use of parameters such as mean annual rainfall and mean annual temperature to differentiate between so-called morphoclimatic regions. This was on the grounds that key geomorphological processes were often the response to complex environmental controls, such that, for example, pluvial erosion is often more closely correlated to individual storm characteristics, including rainfall intensity, duration, and frequency than the total sum of rainfall within the year. In the specific case of weathering, it is obvious that mechanisms such as freeze thaw are highly dependent upon the crossing of specific meteorological thresholds. It is evident that conditions experienced at and near the stone/atmosphere interface are markedly different from those measured within a Stevenson's Screen and can vary considerably, not only within regions but over distances measured in as little as centimeters.

Intraregional variations in stone decay have been extensively demonstrated across Europe through numerous field trials in which samples of selected stone types are exposed under similar controlled conditions for set periods at sites chosen to reflect climatic and/or pollution differences. Typically, trials emphasize differences between rural, maritime, urban, and industrial locations, although they can be used to examine environmental gradients. For example, Smith et al. (1994b) used an exposure trial to relate stone decay in the industrial complex of Marghera to that experienced in the downwind city of Venice. Increasingly, however, trials are designed to examine specific processes. Stone types are selected to be sensitive to the process under consideration and to act as "environmental sensors" rather than as passive collectors of pollutants that happen to land on them (Bidner et al., 2002). In this role they can also be used to test differences in surface modification related to microclimatic variations (e.g., Turkington et al., 1997).

Evaluation of intraregional variations in weathering and erosion has not depended just on specially created exposure trials; many assessments have taken advantage of dated buildings and structures. Halsey et al. (1996), for example, used churches built of ferruginous sandstone in the English West Midlands to map soiling and subsequent decay of the stone and showed how decay was correlated with patterns of urbanization and combustion of coal as a fuel. The structures most commonly used to map weathering and erosion rates are gravestones because of their widespread occurrence and the ability, in most cases, to date their emplacement. Recently wide-ranging studies have sought to map weathering across the UK (Cooke et al., 1995) and the United States (Meierding, 1981). It is clear that weathering rates have been strongly influenced by local and regional patterns of air pollution, and these have changed over time (e.g., Meierding 1993). This is not the only problem encountered in the interpretation of weathering on gravestones. Inkpen (1999) identified difficulties in using the height of lead lettering above the stone as a means of estimating surface loss and stressed the need for both care and consistency in experimental design and data analysis to allow the integration of different studies. To these problems must be added uncertainty over any time lapse between the death of the person memorialized and the erection of the gravestone and, more importantly, the nonlinearity of material loss (Roberts, this volume). This is especially important in the weathering of sandstones (e.g., Matthias, 1967), where under certain conditions many years of apparent stability are followed by the effectively instantaneous loss of a large proportion of the stone through contour scaling (McGreevy et al., 1983). Indeed, gravestones and sandstone buildings in general provide excellent demonstrations of the episodic nature of weathering and the way in which rapid loss is linked to the crossing of strength/stress thresholds. Often this is the result of a combination of the gradual weakening of the stone as, for example, fracture networks develop or constituent minerals are chemically weathered, increased internal stress through, for example, salt accumulation, and an extreme external stress such as a severe frost (Smith et al., 1994a).

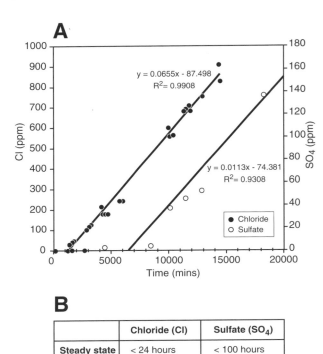

Figure 3. (A) Comparison of effective diffusion coefficients for chloride and sulfate through Dumfries sandstone. (B) Figures for time to reach steady-state and flow rates of chloride and sulfate ions through Dumfries sandstone.

Gravestones not only allow the weathering variations across a region to be plotted, but can also provide insights into microclimatic controls. In particular, thanks to an ecclesiastical concern for east-west orientation, gravestones and churches provide excellent opportunities to examine the effects of aspect on weathering. This is illustrated by the work of Robinson and Williams (1999) on Hastings Beds sandstone gravestones in southeast England, who found that the west facing sides of the gravestones were more weathered. Robinson and Williams (1999) were undecided, however, as to whether this reflected increased wetness from exposure to the prevailing wind and rain or was due to the exposure of the east facing sides to the morning sun, causing those sides to start drying earlier after overnight rain. Their observations have been supported by the work of Meierding (2004) in the northeast United States who found significantly greater contour scaling on the west facing sides of Arkose gravestones.

A true investigation of aspect effects using gravestones is restricted by their normally slab-like shape, which means that they have effectively two faces and have limited scope for the development of cavernous weathering forms such as alveolar weathering and tafoni. Robinson and Williams (1996) demonstrated the importance of these constraints in their study of church towers built of the same Hastings Beds sandstone. In contrast to the gravestones, they found the cool and damp east- and north-facing stonework to be initially more prone to weathering than that facing west and south. Common weathering features included granular disaggregation, cracking, alveolar weathering, and scaling. Case hardening is also found, although its development appears to be inhibited on moist shaded stonework, which might be a factor in its more rapid weathering. The importance of the precise interactions between moisture and thermal regimes as controls on type and rate of weathering has been further demonstrated by the work of Paradise (2002) at the ancient Nabataean city of Petra in Jordan. At a much lower latitude than the previous studies, and under conditions of limited moisture availability, he found both east- and west-facing walls to be more weathered than those facing north and south. This is thought to reflect the fact that eastern and western walls experience complex environmental cycles, including rapid cooling as east-facing walls are thrown into shadow and rapid heating as the afternoon sun reaches west-facing walls. South-facing walls are in the sun all day, experience a more gradual pattern of heating and cooling, and are typically very dry. North-facing walls are in permanent shadow, but in the absence of atmospheric moisture, this appears to impart no weathering advantage.

The preceding discussion of micro-environmental factors has treated them as essentially independent of the weathering processes they control. This is not, however, the general case and it is usual for the weathering environment to be modified as stone is eroded to create complex surface morphologies. This is certainly the case where cavernous forms create distinctive internal microclimates that in hot deserts are typically more humid and cooler than adjacent cliff faces and encourage both direct deposition and absorption of atmospheric moisture (Dragovich, 1981; Smith and McAlister, 1986). The supposition is that hollows may begin simply as a patch of stone that is for some reason more susceptible to weathering. Once a hollow is initiated, positive feedback operates, in which conditions amenable to weathering maintain or even accelerate the growth of the cavern (Turkington, 1998). Eventually, however, the feedback mechanism is self-limiting, perhaps when the cavern is so deep that it is cut off from a regular moisture supply and/or diurnal temperature variations are ameliorated (Turkington and Phillips, 2004). Although this scenario was originally formulated for caverns on rock outcrops, the same reasoning applies to walls in which individual stone blocks retreat, but with the advantage that the lateral boundaries of the hollow are clearly constrained by the surrounding blocks (Smith et al., 2002).

Because of the controls on the morphology of hollows created by block retreat in stone walls, Turkington et al. (2002) were able to devise a series of experiments that allowed monitoring of changing micro-environmental, stone moisture, and temperature conditions as a block retreats. Results from experiments using different thermal regimes, airflow conditions, and wetting of the stone show that under still air conditions, surface rock temperatures actually increased until the block was 50% in shade, after which they fell progressively as the block retreated. However, a forced airflow markedly reduced surface temperatures and sub-surface temperature gradients, aided presumably by evaporation

of absorbed moisture. Turkington et al. (2002) also showed that wetting the stone did not reduce thermal stress on the retreating stone but that it was progressively restricted, together with effective moisture cycling, to the near-surface layer as the block retreated. Stresses resulting from salt accumulation, and other processes controlled by temperature and moisture cycling, are thus concentrated in a shallow layer near the surface of the block, which might go some way to explaining the predominance of multiple flaking noted within many cavernous recesses both on buildings and in nature. Although these findings might begin to explain the detailed patterns of weathering found in cavernous hollows, perhaps the most important aspect of the work is that it demonstrates and begins to quantify the feedbacks that control the mechanisms responsible for decay. It is arguable whether the experiment would have been devised were it not for the commercial imperative of accurately diagnosing the causes of accelerated decay in a local church and identifying appropriate methods for either switching off positive feedbacks or ensuring that they are not switched on in the first place.

CONCLUSIONS

The themes discussed above have been selected as representative of "crossover" areas between studies of natural rock weathering and urban stone decay, which are by no means exhaustive. An extensive body of research into salt attack on sandstone in urban areas has been encapsulated to highlight recent progress in characterizing the operation of, and controls on, complex urban stone decay systems. However, rather than listing further details on research that bridges earth sciences and stone decay studies, it might be more informative to ask why, if there is such an overlap, have earth scientists not shown a greater interest in urban stone decay?

Reasons for the lack of discourse between the two disciplines are many and varied, but have at their heart the simple fact that the research communities have few points of contact. In general, they go to different conferences, publish in different journals, and belong to different professional organizations with specific entry requirements. This polarization has been encouraged by the growth of research assessment exercises for academics in the UK that, for practical and ideological reasons, encourage subject specialization through handing the evaluation process to peer review panels drawn from within rather than across subject areas. Access to stone decay studies is also hindered by the fact that they are frequently to be found as "gray literature" in the form of unpublished reports that can themselves be constrained by client confidentiality. Those studies that do find their way into the wider community often appear in conference proceedings. This in turn can militate against academic participation because of an unhealthy and somewhat incestuous preoccupation with journal impact factors and citation scores and a reliance on international journal articles as measures of success.

The dichotomy between the disciplines also spills over into the remits of research funding agencies. This means that, all too often, interdisciplinary research that would bridge the gap between natural rock weathering and urban stone decay is excluded on the grounds that it does not fall within the remit of the agency. Thus, agencies dealing with the natural environment claim that engineers should fund research with "urban" and "building" in the title, while engineering agencies cannot understand why geologists and/or environmentalists should not foot the bill for weathering studies. As well as these practical and institutional difficulties, there are cultural clashes. For example, stone conservators, the companies that employ them, and the building owners are primarily concerned with rapid and sustainable solutions to decay problems. This is in contrast to a perception of academics as people who delight in the discovery of new problems and have little regard for deadlines. Academics are in turn not immune from stereotyping, and a residual elitism remains that views applied research, on projects that are not peer reviewed, as in some way second class (Smith et al., 2003).

Overcoming these barriers to the exchange of ideas requires a mutual parity of esteem both between academic researchers in both areas and between academics and conservation practitioners. Linked to this is the need to educate earth scientists in the similarities, rather than the differences, between rock weathering and stone decay and in research opportunities that exist within the urban environment; these are certainly more challenging than a simple "weathering experiment." Most importantly, it must be recognized that placing a stone in a building does not immunize it from natural weathering processes. In this context, it is appropriate to leave the last word to Stephen Jay Gould and a question that he raised in one of his last publications (2000, p. 152): "Why should intellectual content correlate with difficulty of physical access—a common supposition that must rank among the silliest of romantic myths? … By all means, take that dogsled across the frozen wastes if no alternative exists, but if the A train also goes to the same destination, why not join Duke Ellington for the smoother ride?" The breadth, depth, and sheer quantity of his publications testify that this is not a call for the quiet life, but instead it draws attention to the benefits of taking notice of what is under our noses. This comment referred specifically to the use by Charles Lyell of erosion on three Roman pillars at the Temple of Serapis in Pozzuoli as a tide gauge to record extensive and gradual change of land and sea levels, possibly one of the very first examples of urban stone decay used to inform a geological debate.

ACKNOWLEDGMENTS

The writers are indebted to Gill Alexander for preparing the diagrams and to Consarc Conservation Ltd. for facilitating the study of St Matthew's Church.

REFERENCES CITED

Amoroso, G.G., and Fassina, V., 1983, Stone decay and conservation. Material Science Monograph 11: Amsterdam, Elsevier, 447 p.

Arnold, A., 1982, Rising damp and saline minerals, *in* Gauri, K.L., and Gwinn, J.A., eds., Fourth International Congress on the Deterioration and Preservation of Stone Objects, Louisville, July 7–9, 1982: Louisville, Kentucky, The University of Louisville, p. 11–28.

Bidner, T., Mirwald, P.W., Recheis, A., and Bruggerhoff, S., 2002, Stone as sensor material for weathering, *in* Prykryl, R., and Viles, H.A., eds., Understanding and managing stone decay: Prague, Karolinum Press, p. 97–112.

Camuffo, D., 1998, Microclimate for cultural heritage: Amsterdam, Elsevier, 415 p.

Common, R., 1966, Slope failure and morphogenetic regions, *in* Dury, G.H., ed., Essays in Geomorphology: London, Heinemann, p. 53–82.

Cooke, R.U., 1979, Laboratory simulation of salt weathering processes in arid environments: Earth Surface Process, v. 4, p. 347–359.

Cooke, R.U., and Gibbs, G.B., 1993, Crumbling heritage?: Swindon, National Power and Powergen, p. 68.

Cooke, R.U., Brunsden, D., Doornkamp, J.C., and Jones, D.K.C., 1982, Urban geomorphology in drylands: Oxford, Oxford University Press, 324 p.

Cooke, R.U., Inkpen, R.J., and Wiggs, G.F.S., 1995, Using gravestones to assess changing rates of weathering in the United Kingdom: Earth Surface Processes and Landforms, v. 20, p. 531–546.

Cooper, T.P., Dowding, P., Lewis, J.O., Mulvin, L., O'Brien, P.F., Olley, J., and O'Daley, G., 1991, Contribution of calcium from limestone and mortar to the decay of granite walling, *in* Baer, N.S., Sabbioni, C., and Sors, A.I., eds., Science, technology and the European cultural heritage: Oxford, Butterworth-Heineman, p. 456–459.

Curran, J.M., and Smith, B.J., 2000, Non-destructive testing: Using probe permeametry in building stone studies, *in* Rammlmair, D., Mederer, J., Oberthur, Th., Heiman, R.B., and Pentinghaus, H., eds., Applied Mineralogy in Research, Economy, Technology, Ecology and Culture: Rotterdam, Balkema, p. 967–970.

Dragovich, D., 1981, Cavern microclimates in relation to preservation of rock art: Studies in Conservation, v. 26, p. 143–149.

Fitzner, B., and Heinrichs, K., 2002, Damage diagnosis of stone monuments— weathering forms, damage categories and damage indices, *in* Prykryl, R., and Viles, H.A., eds., Understanding and managing stone decay: Prague, Karolinum Press, p. 11–58.

Fitzner, B., Heinrichs, K., and Kownatzki, R., 1997, Weathering forms at natural stone monuments—classification, mapping and evaluation: International Journal for Restoration of Buildings and Monuments, v. 3, p. 105–124.

Galan, E., and Zezza, F., editors, 2000, Protection and conservation of the cultural heritage in the Mediterranean cities: Rotterdam, Balkema, 490 p.

Geikie, A., 1880, Rock-weathering as illustrated in Edinburgh church yards: Proceedings, Royal Society, Edinburgh, v. 10, p. 518–532.

Goudie, A.S., 1974, Further experimental investigation of rock weathering by salt and other mechanical processes: Zeitschrift Für Geomorphologie Supplementband, v. 21, p. 1–12.

Goudie, A.S., 1986, Laboratory simulation of 'the wick effect' in salt weathering of rock: Earth Surface Processes and Landforms, v. 11, p. 275–285.

Goudie, A.S., 1993, Salt weathering simulation using a single-immersion technique: Earth Surface Processes and Landforms, v. 18, p. 369–376.

Goudie, A.S., Cooke, R.U., and Evans, I.S., 1970, Experimental investigation of rock weathering by salts: Area, v. 4, p. 42–48.

Gould, S.J., 2000, The lying stones of Marrakech: Penultimate reflections in natural history: New York, Harmony Books, 372 p.

Hall, C., and Toft, W.D., 2002, Water transport in brick, stone and concrete: London, Spon Press, 317 p.

Halsey, D.P., Dews, S.J., Mitchell, D.J., and Harris, F.C., 1996, The black soiling of sandstone buildings in the West Midlands, England: Regional variations and decay mechanisms, *in* Smith, B.J., and Warke, P.A., eds., Processes of urban stone decay: London, Donhead Publishing, p. 53–65.

Inkpen, R.J., 1999, Gravestones: problems and potentials as indicators of historic changes in weathering, *in* Jones, M.S., and Wakefield, R.D., eds., Aspects of stone weathering, decay and conservation: London, Imperial College Press, p. 16–27.

Inkpen, R.J., Petley, D., and Murphy, W., 2004, Durability and rock properties, *in,* Smith, B.J., and Turkington, A.V., eds, Controls and causes of stone decay: London, Donhead Publishing, p. 33–52.

Matthias, G.F., 1967, Weathering rates of Portland arkose tombstones: Journal of Geological Education, v. 15, p. 140–144.

McAlister, J.J., Smith, B.J., and Curran, J.M., 2003, The use of sequential extraction to examine iron and trace metal mobilisation and the case hardening of building sandstone: a preliminary investigation: Microchemical Journal, v. 74, p. 5–18, doi: 10.1016/S0026-265X(02)00043-7.

McGreevy, J.P., Smith, B.J., and McAlister, J.J., 1983, Stone decay in an urban environment, examples from south Belfast: Ulster Journal of Archaeology, p. 167–171.

Meierding, T.C., 1981, Marble tombstone weathering rates: A transect of the United States: Physical Geography, v. 2, p. 1–18.

Meierding, T.C., 1993, Marble tombstone weathering and air pollution in North America: Annals of the Association of American Geographers, v. 83, p. 568–588.

Meierding, T.C., 2004, Arkose (Brownstone) Tombstone Weathering in the Northeastern USA, *in* Smith, B.J., and Turkington, A.V., eds., Controls and causes of stone decay: London, Donhead Publishing, p. 167–198.

Paradise, T.R., 2002, Sandstone weathering and aspect in Petra, Jordan: Zeitschrift für Geomorphologie, v. 46, p. 1–17.

Peltier, L.C., 1950, The geographical cycle in periglacial regions as it is related to climatic geomorphology: Annals of the Association of American Geographers, v. 40, p. 214–236.

Robinson, D.A., and Williams, R.B.G., 1987, Surface crusting of sandstones in southern England and northern France, *in* Gardiner, V., ed., International Geomorphology 1986 Part II: New York, Wiley & Sons, p. 623–635.

Robinson, D.A., and Williams, R.B.G., 1996, An analysis of the weathering of Wealden sandstone churches, *in* Smith, B.J., and Warke, P.A., eds., Processes of urban stone decay: London, Donhead Publishing, p. 133–149.

Robinson, D.A., and Williams, R.B.G., 1999, The weathering of Hastings Beds sandstone gravestones in south east England, *in* Jones, M.S., and Wakefield, R.D., eds., Aspects of stone weathering, decay and conservation: London, Imperial College Press, p. 1–15.

Saiz-Jimenez, C., editor, 2003, Molecular biology and cultural heritage: Rotterdam, Balkema, 300 p.

Shaát, A., Basheer, P.A.M., Long, A., and Montgomery, F., 1994, Reliability of accelerated chloride migration test as a measure of chloride diffusivity in concrete, *in* International Conference on Corrosion and Corrosion Protection of Steel and Concrete, Sheffield, London: London, Academic Press, p. 446–460.

Siegesmund, S., and Weiss, T., 2002, Natural stone, weathering phenomena, conservation strategies and case studies: London, Geological Society Special Publication 205, 518 p.

Smith, B.J., 1994, Weathering processes and forms, *in* Abrahams, A., and Parsons, A. J., eds., Geomorphology of desert environments: London, Chapman and Hall, p. 39–63.

Smith, B.J., 2003, Background controls on urban stone decay: Lessons from natural rock weathering, *in* Brimblecombe, P., ed., Air Pollution Reviews—Vol. 2: The effects of air pollution on the built environment: London, Imperial College Press, p. 31–62.

Smith, B.J., and Kennedy, E.M., 1999, Moisture loss from stone influenced by salt accumulation, *in* Jones, M.S., and Wakefield, R.D., eds., Aspects of stone weathering, decay and conservation: London, Imperial College Press, p. 55–64.

Smith, B.J., and McAlister, J.J., 1986, Observations on the occurrence and origin of salt weathering phenomena near Lake Magadi, southern Kenya: Zeitschrift für Geomorphologie, v. 30, p. 445–460.

Smith, B.J., and McGreevy, J.P., 1983, A simulation study of salt weathering in hot deserts: Geografiska Annaler, v. 65A, p. 127–133.

Smith, B.J., and McGreevy, J.P., 1988, Contour scaling of a sandstone by salt weathering under simulated hot desert conditions: Earth Surface Processes and Landforms, v. 13, p. 697–706.

Smith, B.J., and Turkington, A.V., 2004, Stone decay, its causes and controls: London, Donhead Publishing, 306 p.

Smith, B.J., Whalley, W.B., and Fassina, V., 1988, Elusive solution to monumental decay: New Scientist, v. 1615, p. 49–53.

Smith, B.J., Magee, R.W., and Whalley, W.B., 1994a, Breakdown patterns of quartz sandstone in a polluted urban environment: Belfast, N. Ireland, *in* Robinson, D.A., and Williams, R.B.G., eds., Rock weathering and landform evolution: Chichester, Wiley, p. 131–150.

Smith, B.J., Whalley, W.B., Wright, J., and Fassina, V., 1994b, Short-term surface modification of limestone test samples: Examples from Venice and the surrounding area: The Conservation of Monuments in the Mediterranean Basin, Proceedings of the 3rd International Symposium, Venice, p. 217–226.

Smith, B.J., Basheer, P.A.M., Long, A.E., McAlister, J.J., and Curran, J.M., 2001, Surface modification of building stone: implications for cleaning and replacement: Final Report of EPSRC Project GR/L957739/01, p. 35.

Smith, B.J., Turkington, A.V., Warke, P.A., Basheer, P.A.M., McAlister, J.J., Meneely, J., and Curran, J.M., 2002, Modelling the rapid retreat of building sandstones. A case study from a polluted maritime environment: Geological Society of London Special Publication 205, p. 339–354.

Smith, B.J., Warke, P.A., Turkington, A.V., Curran, J.M., Stelfox, D., and Savage, J., 2003, The resolution of conflicting demands in University/Industry collaboration on stone conservation: A UK perspective, *in* Kozlowski, R., ed., Proceedings of the 5th European Commission Conference on Cultural Heritage Research, Cracow, Institute of Catalysis and Surface Chemistry: Cracow, Polish Academy of Sciences, p. 215–218.

Stoddart, D.R., 1969, Climatic geomorphology: Review and re-assessment: Progress in Geography, v. 1, p. 161–222.

Turkington, A.V., 1998, Cavernous weathering in sandstone: Lessons to be learnt from natural exposure: Quarterly Journal of Engineering Geology, v. 31, p. 375–384.

Turkington, A.V., and Phillips, J.D.P., 2004, Cavernous weathering, dynamical instability and self-organization: Earth Surface Processes and Landforms, v. 29, no. 6, p. 665–675, doi: 10.1002/esp.1060.

Turkington, A.V., and Smith, B.J., 2000, Observations of three-dimensional salt distribution in building sandstone: Earth Surface Processes and Landforms, v. 25, p. 1317–1332, doi: 10.1002/1096-9837(200011)25:12<1317::AID-ESP140>3.0.CO;2-#.

Turkington, A.V., and Smith, B.J., 2004, Interpreting the spatial complexity of decay features on a sandstone wall: St Matthew's Church Belfast, *in* Smith, B.J., and Turkington, A.V., eds, Controls and causes of stone decay: London, Donhead Publishing, p. 149–166.

Turkington, A.V., Smith, B.J., and Whalley, W.B., 1997, Short-term stone surface modification; an example from Venice: Proceedings of the IVth International Symposium on the Conservation of Monuments in the Mediterranean Basin, Rhodes, v. 1, p. 359–372.

Turkington, A.V., Smith, B.J., and Basheer, P.A.M., 2002: The effect of block retreat on sub-surface temperature and moisture conditions in sandstone, *in* Prykryl, R., and Viles, H.A., eds., Understanding and managing stone decay: Prague, Karolinum Press, p. 113–126.

Wakefield, R.D., and Jones, M.S., 1998, An introduction to stone colonizing micro-organisms and biodeterioration of building stone: Quarterly Journal of Engineering Geology, v. 31, p. 301–314.

Warke, P.A., 1996, Inheritance effects in building stone decay, *in* Smith, B.J., and Warke, P.A., eds., Processes of urban stone decay: London, Donhead Publishing, p. 32–43.

Warke, P.A., and Smith, B.J., 2000, Salt distribution in clay-rich weathered sandstone: Earth Surface Processes and Landforms, v. 25, p. 1333–1342, doi: 10.1002/1096-9837(200011)25:12<1333::AID-ESP141>3.0.CO;2-6.

Warke, P.A., Curran, J.M., Turkington, A.V., and Smith, B.J., 2003, Condition assessment for building stone conservation: A staging system approach: Building and Environment, v. 38, p. 1113–1123, doi: 10.1016/S0360-1323(03)00085-4.

Weiss, T., Siegesmund, S., and Fuller, E.R., 2002, Thermal stresses and microcracking in calcite and dolomite marbles finite element modelling: Geological Society of London Special Publication 205, p. 89–102.

Winkler, E.M., 1975, Stone: Properties, durability in man's environment: New York, Springer-Verlag, 230 p.

MANUSCRIPT ACCEPTED BY THE SOCIETY 19 JANUARY 2005

Printed in the USA

Geological Society of America
Special Paper 390
2005

Can stone decay be chaotic?

Heather A. Viles*

School of Geography and the Environment, University of Oxford, Oxford OX1 3TB, UK

ABSTRACT

Despite decades of study of stone decay phenomena and practical stone conservation experience, there are still great gaps in our knowledge, and it is still difficult
to predict how stone decay and soiling will respond to changes in air pollution and
other environmental changes. Nonlinear behavior and chaotic system dynamics have
been recognized in many earth surface systems and are also relevant to stone decay
phenomena. Identification of nonlinear behavior in stone decay systems should help
interpret, model, and manage such systems. Several examples indicate the presence of
nonlinear, and sometimes chaotic, behavior in stone decay systems. A review of sources of nonlinearity in stone decay systems illustrates that nonlinearities are common
and important, with chaotic behavior sometimes resulting. Such findings illustrate the
potential dangers of applying linear damage functions and the need for management
efforts to take seriously nonlinearities in stone decay, due to the often complex and
chaotic responses of deteriorating stonework to environmental changes.

Keywords: nonlinearity, systems, chaos, complexity, stone decay, thresholds, storage
effects, saturation, depletion, feedback, self-organization, material properties, bioprotection, bioweathering, salt weathering, frost weathering, chemical weathering, synergies,
hysteresis, limestone, sandstone, tafoni, alveoli, dissolution, case hardening, pollution.

WHY SHOULD WE ASK IF STONE DECAY IS CHAOTIC?

It is important to realize that the available data on stone loss represent only the quasi-steady mechanisms of loss, and that it is
not possible to recommend any one statistically based damage
function for general use in predicting rates of stone deterioration.
(Lipfert, 1989, p. 416)

Building stone decay and soiling are costly problems that
affect irreplaceable elements of the world's cultural heritage
(such as the Taj Mahal, India, and the rock-hewn temples of
Petra in Jordan) as well as more modest buildings and structures
worldwide. After more than 100 years of scientific studies of the
nature and causes of stone decay, many fundamental questions

still remain unanswered, and it has proved difficult to produce
simple dose:response functions that could be used to predict how
stone decay rates will change as a function of changing environmental conditions (such as increased or decreased air pollution
levels or climate change). For example, Lipfert (1989) and Butlin
et al. (1992) have produced linear damage functions relating
rainfall amount and pH and air pollution to limestone decay (in
terms of surface loss or weight loss), based on exposure trials and
experimental studies, but their explanatory power is limited, as
noted in the introductory quote. The question arises as to whether
the failure to tackle these basic issues reflects a lack of effort on
the part of stone decay scientists or whether it reveals something
fundamental about the behavior of stone decay that might be better tackled by a fresh approach to the problem. Similar problems
of explanation and prediction have been found in many other
areas of science, especially earth sciences such as geology and
geomorphology, and in these cases nonlinear dynamical systems

*E-mail: heather.viles@geog.ox.ac.uk

Viles, H.A., 2005, Can stone decay be chaotic? *in* Turkington, A.V., ed., Stone decay in the architectural environment: Geological Society of America Special Paper
390, p. 11–16, doi: 10.1130/2005.2390(02). For permission to copy, contact editing@geosociety.org. ©2005 Geological Society of America.

ideas have proved to be helpful in stimulating new ideas and providing plausible explanations for complex features (Bak, 1997; Phillips, 1999; Sivakumar, 2000). Concepts already developed in, or used by, stone decay science may be usefully reevaluated within the context of nonlinear behavior. Modeling of stone decay might also fruitfully examine nonlinear behavior. Identification of nonlinear behavior has implications for management, as it may no longer be possible for a single "stable" state to be recognized and managed. While by no means do all nonlinear systems behave chaotically, some do under certain conditions. Thus, it seems timely and potentially helpful to examine the utility of nonlinear dynamical systems ideas to stone decay science. If we can identify nonlinear stone decay systems and examine their propensity to behave chaotically, then we may be led to alter our approach to prediction and management.

NONLINEAR DYNAMICAL SYSTEMS

Ideas about nonlinear, chaotic or complex system behavior are widespread in science, and often come packaged under headings such as "chaos theory," "complexity theory," or "self-organized criticality," which promise meta-explanations of whole swathes of phenomena. Such different approaches also often present differing explanations of basic characteristics, and there is also often a conflict between the mathematical formalism involved in analyzing nonlinear systems and the scientific conceptualization by earth scientists and others of the systems under study. All these issues make it hard for many scientists to understand the importance and relevance of a nonlinear dynamical systems approach to their area of study and in turn make it difficult for nonlinear mathematicians to understand the nature of the environmental questions being asked.

Despite the risk of presenting an oversimplified caricature of nonlinear dynamical systems ideas, we can identify certain characteristics. Nonlinear systems are those in which outputs do not change proportionally to a change in inputs over the entire range of inputs. To give an example relevant to building stone decay, increases or decreases in air pollution (or other environmental parameters) do not necessarily lead to proportional increases or decreases in surface lowering rates even under controlled experimental conditions (Guidobaldi and Mecchi, 1993). Nonlinear systems are common in the world around us, and their behavior may often be simple. However, many nonlinear systems have been found to show chaotic behavior patterns at some times. Seemingly random and unpredictable behavior (or chaos) can thus emerge from deterministic systems. Certain general characteristics of such systems prone to chaotic behavior have been identified, such as sensitive dependence on initial conditions, as well as the potential for self-organizing behavior and unpredictability. If stone decay systems have these characteristics, then we might expect chaotic behavior to occur.

Sensitive dependence on initial conditions is often found in systems that contain multiple elements and nonlinear interactions, especially when they are forced and dissipative (Baas, 2002). In soil science, for example, chaotic behavior has been identified from

high variability over space and time of soil profiles (or outputs) in many areas that cannot be explained by accompanying variations in the major soil-forming factors (or inputs), as reported by Phillips (1993). Identification of such behavior does not mean that all is unpredictable; rather, it may help explain under what conditions predictions can be made and what sort of features we can predict.

NONLINEAR DECAY PROCESSES

There are many examples of nonlinear processes involved in stone decay, which might produce nonlinear behavior in overall weathering systems. For example, several chemical weathering processes are described by exponential or logarithmic rates of the general form:

$$d = kt^{1/n} \qquad (1)$$

$$d = k \log (t), \qquad (2)$$

where $n > 1$, k = constant, t = time, and d = decay.

Indeed, Colman (1981) showed that most weathering process rates, apart from congruent solution, decrease over time—at least over the hundreds to thousands of year time spans relevant to most geomorphological situations. Support for this suggestion comes from a wide range of field settings. For example, Matsukura and Matsuoka (1991) found that the development of tafoni on dated coastal terrace surfaces in Japan could be approximated by an exponential function:

$$D = 20.3 \, (1 - e^{-0.005t}), \qquad (3)$$

where D = mean of 10 largest tafoni depths in cm and t = time in years.

Over timescales more relevant to stone decay, Lipfert (1989, p. 416) writes that "Skoulikidis (1982) showed in laboratory experiments a parabolic relationship of gypsum formation with time, which is consistent with SO_2 deposition being controlled by diffusion through corrosion products." Hoke and Turcotte (2002, 2004) utilize an exponential equation, similar to that in Equation 1, to model the developing dissolution layer on a weathering surface based on a diffusion-dominated process. Weathering of porous building materials is most likely to be a diffusion-based process, as it is not simply a surface phenomenon, and penetration of water and reactive compounds into the near-surface layer is an integral part of the decay process. Overall, there has been much debate over what the causes of such nonlinear weathering rates might be and how important they might be to weathering over the tens to hundreds of year time spans of most relevance to most stone decay.

IDENTIFYING NONLINEARITY AND CHAOS IN STONE DECAY

So how can we tell if nonlinear dynamical systems behavior is present in stone decay? Two simple approaches are used here:

first, to conceptualize stone decay in terms of a nonlinear dynamical system and find examples of potentially chaotic behavior, and second, to identify sources of nonlinearity in stone decay systems.

As Phillips (2003, p. 4) pointed out, "In applying nonlinearity to earth surface systems, the terms 'input' and 'output' may need to be more broadly and flexibly defined than their mathematical or systems theory meanings." This is equally true of attempts to understand stone decay in terms of nonlinear systems behavior. If we can conceive of the "stone decay system" as involving a series of inputs such as precipitation, pollutants, physical stresses, and microorganisms, then the outputs could be thought of as the visible signs of decay such as blistering, loss of material, or cracking. One of the problems with much stone decay research has been the difficulty of providing meaningful quantifications of inputs and, especially, outputs. In most cases, this simple "stone decay system" has a series of manifestations at different scales (from individual mineral grains through individual stone blocks to whole walls) and is clearly a multidimensional system (involving a range of interlinked chemical, physical, and biological processes). It is certainly reasonable given this basic conceptualization of the "stone decay system" to represent it as a deterministic system in which system behavior should be able to be represented by a series of equations. However, elements of random behavior may also be present and may confound any attempts to identify deterministic chaos. The lack of good data sets on inputs and outputs makes it difficult to interpret the system behavior in any detailed way and thus makes it difficult to prove the existence of chaotic behavior.

Despite the caveats listed above, we can identify some possible manifestations of chaotic or complex behavior in the simple "stone decay system"—three of which are discussed here. First, observations of rhythmical weathering phenomena, such as alveoli, have been linked to self-organizing behavior in nonlinear systems (Mikulas, 2001; Turkington and Phillips, 2004). Such alveoli are often found on walls subjected to salt weathering and although several explanations have been proposed, no general theory of their formation exists (Goudie and Viles, 1997). Second, the "memory effect," as identified by Cooke in BERG (Building Effects Review Group, 1989), whereby building stone blocks may continue to weather at a high rate even after pollution has ameliorated because of the reactivation of pollutants stored within the blocks, may also be a manifestation of nonlinear behavior. Phillips (1999, p. 121), discussing similar phenomena in soils, sees them as a manifestation of dynamical instability, chaos, and self-organization within the soil system. Third, the observation of much seemingly complex behavior in some simple stone decay systems may also reveal the presence of nonlinearity. For example, exposure trials often produce very complex data sets, in which simple relationships between pollution levels and/ or rainfall can be very hard to detect. Studies such as the UK National Materials Exposure Programme (NMEP), for example, found some very complex patterns of decay and soiling on standard stone tablets exposed in a range of different environments (Viles et al., 2002). Replicate tablets often showed very different patterns of soiling, weight loss, and uptake of pollutants. Although these findings could also be a result of poor experimental design, randomness, or the inherent complexity of the stone decay system operating on these tablets, the presence of nonlinearity producing complex results is a reasonable hypothesis, which justifies further examination. Similarly, the complex temporal trends in surface lowering on the balustrade wall of St Paul's Cathedral, London, over a 20-year time span may be viewed as nonlinear outputs of the time series of precipitation and sulfur dioxide inputs, which may explain the lack of simple correlation between the various data series (Trudgill et al., 2001).

The three examples presented above illustrate that there are plausible examples of complex nonlinear behavior to be found within a simply conceptualized "stone decay system." The complex behavior observed in both the NMEP and St Paul's Cathedral studies appears to be a classic case of divergence in the absence of any observable variation in inputs or environmental controls, which results from dynamical instability and chaos. We should not forget, however, that there are other potential causes of such complex behavior and also that some seemingly complex behavior may actually mask more simple trends that we have yet to unearth. Stone decay scientists are now much better positioned to be able to interpret system behavior and prove the existence of chaotic behavior because of the development of improved data sets and data collection techniques. The use of laser scanning, for example, allows accurate, high-resolution repeat measurements of the morphology of large areas of building surfaces. Many more building and monument surfaces are now instrumented, and there is now a range of long-term data sets from exposure trial studies.

POTENTIAL SOURCES OF NONLINEARITY IN STONE DECAY

Another approach to discovering whether nonlinearity might be an important issue for stone decay science is to identify possible sources of nonlinearity within stone decay systems. Phillips (2003) provided a useful template for geomorphology, which is simplified and adapted here for stone decay science (see Table 1). Ten different causes of nonlinearity are discussed; i.e., thresholds, storage effects, saturation and depletion, self-reinforcing processes, self-limiting processes, competitive interactions, self-organization, synergisms of processes, initial relief, and hysteresis. As Phillips (2003) acknowledged, there are many semantic and conceptual difficulties involved with untangling the behavior of complex earth surface systems, and thus one person's idea of a source of nonlinearity might easily be viewed by someone else as a type of nonlinear response. Furthermore, many sources of nonlinearity may be interrelated and operating together, and some are hard to categorize. However, it is worth attempting to identify and separate sources of nonlinearity, as they might provide the key to understanding what is going on and thus aid our attempts to predict and manage.

Thresholds have been commonly thought of as playing an important role in many weathering processes. In basic terms, we

TABLE 1. SOURCES OF NONLINEARITY IN STONE DECAY SYSTEMS

Source	Why is it nonlinear?	Examples
Thresholds	Outputs will not be proportional to inputs both above and below a threshold.	Critical stress needed to be exceeded to cause deterioration
Storage effects	The addition or removal of stored components will affect input: output relationships.	The memory effect
Saturation and depletion	The effect of an increase in input varies with respect to some optimum.	Weathering rind development
Self-reinforcing, positive feedback	The action of a process accelerates without any change in input rate.	Tafoni development
Self-limiting processes	The action of a process decelerates without any change in input rate.	Case hardening
Competitive interactions	The balance between two competing processes can be catastrophically affected by changes in inputs.	Competing processes of bioprotection and biodeterioration
Self-organization	A system may show complex adaptations without changes in inputs.	Alveoli development
Synergies	Two processes operating together can magnify the effects of inputs.	Frost and salt weathering
Material/building properties	The initial characteristics (either inherent or created by building) may amplify or retard the effect of inputs.	Stone carving creating weathering "hotspots"
Hysteresis	There may be two or more possible values of an output for a given input.	Cyclic processes leading to fatigue

can see that inputs will lead to one type of outcome below a threshold and another type of outcome above that threshold. Thresholds thus, by definition, indicate nonlinearity. For example, in physical weathering processes involving the imposition of stress in porous stones (such as freeze-thaw or salt crystallization), there is no measurable weathering output until the stress exceeds the strength of the stone; beyond this threshold, cracking and disintegration may occur at a fast rate. Similarly, Warke and Smith (2000) presented salt weathering in clay-rich sandstone as a threshold phenomenon where disintegration only occurs when a stress/strength threshold is crossed. Some attempts have been made to quantify such thresholds. For example, Paradise (1995) identified two potentially important weathering thresholds in his study of weathering rates in the Roman Theatre at Petra, Jordan, related to iron contents of the rock and receipt of insolation, with abrupt changes in rate occurring around 2% iron concentration and 5200–5300 megajoules m^{-2}.

Storage effects relate to situations in which the addition or removal of stored components affects input:output relationships. For example, the "memory effect" can be viewed as a storage effect, as depending on the storage of pollutants within the stone, that stone will react in a variety of ways to its current environmental conditions. The reality of the memory effect has been tested experimentally by Vleugels et al. (1993), who found no evidence of it, and through modeling by Antill and Viles (2003), who found it likely to occur under certain circumstances. A further storage effect may be the "incubation time" recognized by Cooke et al. (1995), and others, when studying the response of gravestones to polluted atmospheres, and modeled by Hoke and Turcotte (2002). In this case, addition of pollutants to the stone "store" appears to be required before any surface lowering occurs.

Saturation and depletion effects occur when the effects of an increase or decrease in an input varies with respect to some optimum. For example, in biological systems, the law of limiting factors affecting plant growth can be seen in terms of saturation and depletion. In weathering systems, like many biological systems, there is a host of inputs likely to have an effect at the same time, and at some stages saturation may limit the impact of one or more of these. Saturation and depletion may be a good way of conceptualizing the time-dependent interaction between surface lowering and development of a weathering rind or crust. Depletion effects may be evident on the balustrade of St Paul's Cathedral in London, where declining rainfall acidity has recently been accompanied by a change in the nature of surface lowering observed using the micro-erosion meter from overall lowering to patchy lowering, interspersed with stasis or even deposition (Trudgill et al., 2001).

Self-reinforcing positive feedback occurs where internal changes within a system enhance the outputs, beyond any simple relationship with inputs. In terms of weathering, a simple example would be the production of higher surface roughness through weathering processes, which then amplifies the subsequent rate of weathering. Such positive feedbacks have often been proposed as partial explanations for the development of features such as tafoni (e.g., Turkington et al., 2002; Turkington and Phillips, 2004). In the case of tafoni, once a hollow is initiated, the conditions developing within that hollow become increasingly favorable for hollow development, producing an acceleration of hollow formation. Other positive feedbacks occur in phenomena such as deposition of pollutant particles on stone surfaces, where initiation of deposition causes the surface to become more irregular, thus attracting faster subsequent deposition. Microorganism colonization may operate in a similar way, as initial biofilms exude sticky substances that then encourage further colonization.

Self-limiting, or negative feedback, processes are also nonlinear and occur where an increase in inputs leads to declining outputs. In many cases, such self-limiting processes lead to stability, but in some instances they can induce complex behavior.

There are many examples of self-limiting processes within stone decay. For example, the production of a case-hardened surface layer as a result of movement of pore fluids to the surface can limit the subsequent rates of weathering of that surface. A clear example is shown on the Madara Horseman sculpture carved into a sandy limestone cliff in NE Bulgaria. Here, a hard carbonate crust is developing that is producing a state of "auto-conservation" on some parts of the monument surface (Delalieux et al., 2001). At a smaller scale, the observations of "self-passivation" of calcite surfaces after exposure to urban conditions for 6–12 months leading to a lowering of reactivity (and thus a decrease in the likely dissolution rate) by Wilkins et al. (2001) indicate similar self-limiting effects. The depletion of weatherable minerals provides another good example of such self-limiting behavior.

Competitive interactions may also induce nonlinear behavior, when the competition between two processes causes systems to change abruptly. Competition between vegetation growth and erosion in semiarid environments, for example, leads to complex, unstable behavior with disturbance inducing erosion in some circumstances and vegetation growth in others (Thornes, 1990). In stone decay systems, the competition between lichen growth (associated with surface stabilization) and salt weathering (leading to rapid surface disintegration) may provide similar sources of nonlinear, unstable behavior. A slight alteration in environmental conditions might trigger a dramatic change to the domination of lichen colonization or salt weathering. In terms of lichen impacts on stone themselves, we might usefully conceptualize a competition between bioweathering by the lichen (through biochemical and biophysical breakdown under the thallus) and bioprotection (through the net protection afforded by the lichen covering the stone surface). A slight change in environmental conditions, or indeed death of the lichen through natural aging, may induce a shift to bioweathering over bioprotection. Kumar and Kumar (1999) review some of the potential competitive interactions between the various agents of biodeterioration.

Self-organizing behavior involves complex adaptations independent of changing inputs and is commonly found in environmental systems. For example, channeled flow and rills may form as a result of self-organization. Such self-organization may occur where systems are evolving in order to maximize energy dissipation and equalize energy expenditure throughout the system (Phillips, 2003). As mentioned earlier, the rhythmical topography of alveoli might conceivably be a product of self-organization (although there is little agreement on what processes may be involved). Furthermore, water flow over and into porous rocks may also behave in self-organizing ways and thus have a knock-on impact on the pattern of decay features.

Synergisms between processes may also produce nonlinear outcomes as the effect of the two processes together is greater than the sum of the two acting individually. Several studies have illustrated such synergisms under experimental conditions. Papida et al. (2000), for example, identify faster rates of physical weathering on test specimens inoculated with microorganisms than on fresh specimens under similar experimental conditions. They pro-pose that bacterial populations contribute to damage directly and indirectly. Williams and Robinson (1981) found similar synergies between salt and frost weathering on sandstone in laboratory experiments, with salts notably accelerating the decay process. Indeed, it has been proposed that such synergistic co-associations of decay processes (along with antagonistic ones, whereby the actions of two processes together may diminish the overall weathering impact) represent a significant gap in our knowledge of, and ability to predict, deterioration rates and should be the focus of future research efforts (Koestler et al., 1994).

The initial properties of a stone or building surface also have the potential to induce nonlinear behavior in the decay system. Often these take the form of material preparation or architectural factors but can also reflect inherent material properties. For example, one putative cause of the "incubation period" in gravestone weathering whereby no visible decay occurs for several years is the surface polish applied by monumental masons. Similarly, the complex geometrical surfaces created by the carving of stone artifacts may influence the subsequent operation of decay processes. Both these examples illustrate the sensitive dependence on initial conditions of many stone decay systems, a key characteristic of many systems prone to chaotic behavior. A further source of nonlinear behavior in many buildings is the juxtaposition of a range of materials, such as different stone types, bricks, and mortar. Finally, an example of inherent material properties causing nonlinear behavior is the anisotropic thermal expansion behavior of calcite crystals, which can produce stresses in marbles subject to heating and cooling cycles, causing damage (Weiss et al., 2003).

Hysteresis occurs when there are two possible values of an output for a given input. Such situations may be common in many weathering systems when, for example, the upward limb of a heating or wetting cycle may induce different effects to the falling limb, ultimately producing fatigue. Snethlage and Wendler (1997) demonstrated this effect when they showed that hygric dilatation of sandstone contaminated with NaCl under wetting-drying cycles produced irreversible dilatation, increasing with each cycle.

This review of ten potential sources of nonlinearity in stone decay presents many plausible examples, many of which are highly interrelated and most of which require further study in order to ascertain whether they can produce chaotic behavior.

THE WAY FORWARD?

As shown in the preceding sections, nonlinearities can be found in many aspects of stone decay. Some have been intensively studied (such as the exponential growth of chemical weathering rinds), but in many cases, linear approximations have been used instead. The presence of important, and/or multiple, sources of nonlinearity in any single stone decay system may mean that chaotic behavior is likely to occur. A key goal for stone decay scientists now is to provide more data on stone decay system inputs and outputs in order to identify the presence of such nonlinear behavior and to elucidate how likely it is to produce chaotic outcomes.

Failure to recognize such complex behavior will lead to erroneous management strategies in the face of environmental change. For example, building conservators are often faced with the challenge of patchily deteriorated façades in which some blocks are seriously damaged and in urgent need of repair while others are sound. Such patchy deterioration may be the result of initial weaknesses in specific blocks of stone or the result of variations in microclimatic conditions, but it may also reflect the emergence of decay "hotspots" as a result of nonlinear, chaotic behavior. If the patchiness can be explained by initial weaknesses or microclimate variations, then prediction of the trends of decay over time can be made relatively easily. If, however, chaotic conditions apply, then decay "hotspots" will occur spontaneously, and probabilistic prediction would be more appropriate. The key for effective assessment and management of risk is to be able to recognize which is the most plausible explanation under different circumstances and develop appropriate predictive strategies.

ACKNOWLEDGMENTS

I thank Bellie Sivakumar and Jonathan Phillips for their constructive comments on an earlier draft, as well as the many stone decay scientists with whom I have discussed these issues, and the chaos discussion group in the School of Geography and the Environment, University of Oxford, for stimulating my thoughts.

REFERENCES CITED

Antill, S.J., and Viles, H.A., 2003, Examples of the use of computer simulation as a tool for stone weathering research: Building and Environment, v. 38, p. 1243–1250, doi: 10.1016/S0360-1323(03)00081-7.

Baas, A.C.W., 2002, Chaos, fractals and self-organization in coastal geomorphology: Simulating dune landscapes in vegetated environments: Geomorphology, v. 48, p. 309–328, doi: 10.1016/S0169-555X(02)00187-3.

Bak, P., 1997, How nature works: The science of self-organized criticality: Oxford, Oxford University Press, 212 p.

Building Effects Review Group (BERG), 1989, The effects of acid deposition on buildings and building materials in the United Kingdom: London, Her Majesty's Stationery Office, 106 p.

Butlin, R.N., Coote, A.T., Devenish, M., Hughes, I.S.C., Hutchens, C.M., Irwin, J.G., Lloyd, G.O., Massey, S.W., Webb, A.H., and Yates, T.J.S., 1992, Preliminary results from the analysis of stone tablets from the National Materials Exposure Programme (NMEP): Atmospheric Environment, v. 26B, p. 189–198.

Colman, S.M., 1981, Rock weathering rates as a function of time: Quaternary Research, v. 15, p. 250–264, doi: 10.1016/0033-5894(81)90029-6.

Cooke, R.U., Inkpen, R.J., and Wiggs, G.F.S., 1995, Using gravestones to assess changing rates of weathering in the United Kingdom: Earth Surface Processes and Landforms, v. 20, p. 531–546.

Delalieux, F., Cardell, C., Todorov, V., Dekov, V., and Van Grieken, R., 2001, Environmental conditions controlling the chemical weathering of the Madara horseman monument, NE Bulgaria: Journal of Cultural Heritage, v. 2, p. 43–54, doi: 10.1016/S1296-2074(01)01105-0.

Goudie, A.S., and Viles, H.A., 1997, Salt weathering hazards: Chichester: John Wiley, 241 p.

Guidobaldi, F., and Mecchi, A.M., 1993, Corrosion of ancient marble monuments by rain: Evaluation of pre-industrial recession rates by laboratory simulations: Atmospheric Environment, v. 27B, p. 339–351.

Hoke, G.D., and Turcotte, D.L., 2002, Weathering and damage: Journal of Geophysical Research B: Solid Earth, v. 107, no. 10, p. 1–6.

Hoke, G.D., and Turcotte, D.L., 2004, The weathering of stones due to dissolution: Environmental Geology, v. 46, p. 305–310, doi: 10.1007/s00254-004-l033-0.

Koestler, R.J., Brimblecombe, P., Camuffo, D., Ginell, W.S., Graedel, T.E., Leavengood, P., Petushkova, J., Steiger, M., Urzi, C., Verges-Belmin, V., and

Warscheid, T., 1994, Group report: How do external environmental factors accelerate change?, in Krumbein, W.E., Brimblecombe, P., Cosgrove, D.E., and Stainforth, F., eds., Durability and change: The science, responsibility and cost of sustaining cultural heritage: Chichester, Wiley, p. 149–163.

Kumar, R., and Kumar, A.V., 1999, Biodeterioration of stone in tropical environments: An overview: Research in Conservation Series, Los Angeles, The Getty Conservation Institute, 85 p.

Lipfert, W.T., 1989, Atmospheric damage to calcareous stones: comparison and reconciliation of recent experimental findings: Atmospheric Environment, v. 23, p. 415–429, doi: 10.1016/0004-6981(89)90587-8.

Matsukura, Y., and Matsuoka, N., 1991, Rates of tafoni weathering on uplifted shore platforms in Nojima-zaki, Boso Peninsula, Japan: Earth Surface Processes and Landforms, v. 16, p. 51–56.

Mikulas, R., 2001, Gravity and orientated pressure as factors controlling 'honeycomb weathering' of the Cretaceous castellated sandstones (northern Bohemia, Czech Republic): Bulletin of the Czech Geological Society, v. 76, p. 217–226.

Papida, S., Murphy, W., and May, E., 2000, Enhancement of physical weathering of building stones by microbial populations: International Biodeterioration & Biodegradation, v. 46, p. 305–317, doi: 10.1016/S0964-8305(00)00102-5.

Paradise, T.R., 1995, Sandstone weathering thresholds in Petra, Jordan: Physical Geography, v. 16, p. 205–222.

Phillips, J.D., 1993, Chaotic evolution of some coastal plain soils: Physical Geography, v. 14, p. 566–580.

Phillips, J.D., 1999, Earth surface systems: Complexity order and scale: Oxford, Blackwell, 180 p.

Phillips, J.D., 2003, Sources of non-linearity and complexity in geomorphic systems: Progress in Physical Geography, v. 27, p. 1–23, doi: 10.1191/0309133303pp340ra.

Sivakumar, B., 2000, Chaos theory in hydrology: important issues and interpretations: Journal of Hydrology, v. 227, p. 1–20, doi: 10.1016/S0022-1694(99)00186-9.

Skoulikidis, T.N., 1982, Atmospheric corrosion of concrete reinforcements, limestones and marbles, in Ailor, W.H., ed., Atmospheric corrosion: Chichester, John Wiley, p. 807–825.

Snethlage, R., and Wendler, E., 1997, Moisture cycles and sandstone degradation, in Baer, N.S., and Snethlage, R., eds., Saving our architectural heritage: Chichester, John Wiley, p. 7–24.

Thornes, J.B., 1990, The interaction of erosional and vegetation dynamics in land degradation: Spatial outcomes, in Thornes, J.B., ed., Vegetation and erosion: Chichester, John Wiley, p. 41–53.

Trudgill, S.T., Viles, H.A., Inkpen, R., Moses, C., Gosling, W., Yates, T., Collier, P., Smith, D.I., and Cooke, R.U., 2001, Twenty-year weathering remeasurements at St Paul's Cathedral, London: Earth Surface Processes and Landforms, v. 26, p. 1129–1142, doi: 10.1002/esp.260.

Turkington, A.V., and Phillips, J.D., 2004, Cavernous weathering, dynamical instability and self-organization: Earth Surface Processes and Landforms, v. 29, p. 665–675, doi: 10.1002/esp.1060.

Turkington, A.V., Smith, B.J., and Basheer, P.A.M., 2002, Sub-surface temperatures and moisture changes in sandstone, in Prikryl, R., and Viles, H.A., eds., Understanding and managing stone decay: Prague, The Karolinum Press, p. 113–126.

Viles, H.A., Taylor, M.P., Yates, T.J.S., and Massey, S.W., 2002, Soiling and decay of NMEP limestone tablets: The Science of the Total Environment, v. 292, p. 215–229, doi: 10.1016/S0048-9697(01)01124-X.

Vleugels, G., Dewolfs, R., and Van Griecken, R., 1993, On the memory effect of limestone for air pollution: Atmospheric Environment, v. 27A, p. 1931–1934.

Warke, P.A., and Smith, B.J., 2000, Salt distribution in clay-rich weathered sandstone: Earth Surface Processes and Landforms, v. 25, p. 1333–1341, doi: 10.1002/1096-9837(200011)25:12<1333::AID-ESP141>3.0.CO;2-6.

Weiss, T., Siegesmund, S., and Fuller, E.R., 2003, Thermal degradation of marble: indications from finite element modelling: Building and Environment, v. 38, p. 1251–1260, doi: 10.1016/S0360-1323(03)00082-9.

Wilkins, S.J., Compton, R.G., Taylor, M.P., and Viles, H.A., 2001, Channel flow cell studies of the inhibiting action of gypsum on the dissolution kinetics of calcite: a laboratory approach with implications for field monitoring: Journal of Colloid and Interface Science, v. 236, p. 354–361, doi: 10.1006/jcis.2000.7418.

Williams, R.B.G., and Robinson, D.A., 1981, Weathering of sandstone by the combined action of frost and salt: Earth Surface Processes and Landforms, v. 6, p. 1–9.

MANUSCRIPT ACCEPTED BY THE SOCIETY 19 JANUARY 2005

Geological Society of America
Special Paper 390
2005

Weathering of serpentine stone buildings in the Philadelphia, Pennsylvania, region: A geographic approach related to acidic deposition

Thomas C. Meierding

Department of Geography and Center for Climatic Research, University of Delaware, Newark, Delaware 19716, USA

ABSTRACT

Deteriorated stones in urban buildings are often anecdotally cited as evidence of anthropogenic acidic deposition, but that hypothesis has seldom been tested by quantitative urban to rural surface loss comparisons. In the Philadelphia region, semiquantitative measurements (mean maximum pit depths and areal percent surface exfoliation) from all suitable nineteenth century serpentine structures (38 building sides, 6000 blocks) suggest that serpentine damage is greatest where sulfur-based acidic deposition (indicated by nearby marble gravestones) was also maximized. However, inter- and intra-building variations, such as serpentine block composition, groundwater wicking, building geometry, surface aspect (east and south walls are most weathered), and chemical reactions between serpentine and adjacent carbonate-rich materials all have an impact and statistically weaken the acidic deposition–serpentine weathering relationship.

Keywords: weathering, serpentine, acid deposition, surface recession, acid precipitation, quantification, urban-rural differences, pit depth, Pennsylvania, aspect, groundwater wicking, stone replacement, repair, rock properties.

INTRODUCTION

Surfaces of many historic and aesthetically pleasing stone buildings in cities have visibly deteriorated, leading to expensive repairs or to building demolition. In the many studies of individual stone buildings, enumerated by Charola (1998), it has become common to ascribe much of the urban building stone damage to present or past concentrations of sulfur dioxide (Amaroso and Fassina, 1983; Winkler, 1994) or to acid rain (e.g., Charola, 1988; Price, 1996).

However, quantitative, or even qualitative, comparisons of the severity of building stone degradation between urban and rural structures containing similar stone type have rarely been made. An important exception is the work of Haber et al. (1988) in Cracow,

Poland, and its surrounding countryside. Their SEM-EDX (scanning electron microscopy–energy dispersive X-ray), X-ray diffraction, and thin-section analyses made it clear that weathering of limestone structures is greater in the city, where presently there are greater measured SO_2 and NO_x concentrations.

A few studies of tombstones have also geographically tested the air pollution–stone weathering hypothesis. Large urban to rural differences in gravestone surface loss have been measured for limestone (Küpper, 1975) and marble (Baer and Berman, 1983; Feddema and Meierding, 1987; Meierding, 1993; Schreiber and Meierding, 1999). By contrast, Portland arkose sandstone monuments ("brownstones"), while often quite weathered, have proven no more so in polluted than in unpolluted sites (Meierding, 2004).

Meierding, T.C., 2005, Weathering of serpentine stone buildings in the Philadelphia, Pennsylvania, region: A geographic approach related to acidic deposition, *in* Turkington, A.V., ed., Stone decay in the architectural environment: Geological Society of America Special Paper 390, p. 17–25, doi: 10.1130/2005.2390(03). For permission to copy, contact editing@geosociety.org. ©2005 Geological Society of America.

The lack of evidence of a clear relationship between acidic deposition and brownstone weathering demonstrates that each stone type must be evaluated separately for effects of atmospheric acidity, and urban–rural comparisons provide a relatively rapid and economical way to do this. In this case study, serpentine block buildings located throughout the greater Philadelphia region are analyzed semi-quantitatively to gauge the possible effects, or lack thereof, of acidic deposition on a single (by name) stone type. Measured rates of marble tombstone weathering in nearby cemeteries are used as an approximation of the long-term acidic deposition dosage to the serpentine buildings.

LONG-TERM ACIDIC DEPOSITION IN THE PHILADELPHIA REGION

Sulfur Dioxide Concentrations

As with other cities, historic Philadelphia gaseous air pollution concentrations can only be estimated, but the city undoubtedly has been among the most polluted in North America. A computer model by Lipfert (1987) that incorporated population size, fuel source, and industry type suggests that SO_2 concentrations for central Philadelphia were low ($80 \, \mu g/m^{-3}$) in the 1880s, highest ($500 \, \mu g/m^{-3}$) during the 1930s, and low again ($100 \, \mu g/m^{-3}$) by 1980. The close match between model and field-measured SO_2 concentrations during the latter half of the twentieth century lends credence to the model's earlier estimates.

Crude single-year maps based on measured data (City of Philadelphia, 1983; Mather, 1968) demonstrated that in the later twentieth century, SO_2 and visible particulate concentrations were greatest where high population densities and coal-fired industries coexisted, near central Philadelphia. The non-industrialized countryside had low pollutant levels in the past 30 years and presumably during earlier time periods as well.

Mean surface recession rates of vertical, white Vermont-marble slab gravestones (<0.5 mm calcite grain diameter and no major graphite veins) have been shown to correlate with average, long-term SO_2 gas concentrations in North America (Meierding, 1993). The many (>100) evenly distributed, large cemeteries of the Philadelphia region permit construction of detailed marble tombstone surface recession rate maps (Feddema and Meierding, 1987; Meierding, 1993) that closely resemble spatial patterns established by modern SO_2 measurements. Thus, surface recession rates of vertical Vermont gravestones, which vary by almost a tenfold difference between downtown Philadelphia and rural locations 20 km away, provide an estimate of cumulative SO_2 input dosage to nearby serpentine buildings. In downtown Philadelphia, SO_2 damage to stone is probably augmented by NO_x from heavy automotive traffic, along with high summer water vapor content, according to laboratory studies by Johansson et al. (1988). Those catalytic effects cannot be resolved by data from marble tombstone decay, and the three gases are taken together as "dry acidic deposition" in this report.

Acid Precipitation

For the Philadelphia region, wet acidic deposition for the past 30–40 yr has primarily been delivered by westerly winds from tall stack, coal-fired power plants in the Midwestern United States. The sulphuric and nitric acid precipitation emanating from those sources, while chemically potent, should generally affect urban and rural stones equally in the Philadelphia region.

It has been suggested, however, that cities create their own additional precipitation acidity due to "washout" (rain drops that incorporate locally produced acidic gases and particulates as they fall). Recent evidence from precipitation chemistry monitoring sites indeed suggests that Philadelphia augments the acidity of the precipitation that falls on the city center by a small amount (Sherwood, 1990, p. 112); further, the city was more polluted in the early 1900s and should have generated even more urban acidic deposition in that period.

According to data from ground-level horizontal marble table stones that are not much affected by SO_2 gas but instead act as dosimeters of sulphuric, nitric, and carbonic acid precipitation and possibly acidic particulates (Meierding, 2000), central Philadelphia atmospheric conditions historically and cumulatively augmented marble weathering by approximately doubling the rate observed in rural areas. Thus, both dry (gaseous) and wet (rain, snow) acidic precipitation inputs to serpentine buildings have been greater in central Philadelphia than in the countryside.

SERPENTINE IN THE PHILADELPHIA REGION

Genesis

Serpentine outcrops in a narrow, broken band that extends from Alabama to Newfoundland, paralleling the compression structures in the ancestral Taconic and Allegheny orogens (Dann, 1988; Smith and Barnes, 1998). The rock was formed from hydrothermal alteration of magnesium-rich and iron-rich (olivine-rich dunites, peridotites, pyroxenites) seafloor rocks or magmas underneath island arcs. Those rocks were transported and accreted onto the continent before Permian continental collision of North America and Afro-Europe.

Description

Technically, the rock in this report is serpentinite, but following common usage, it is referred to here as "serpentine," a stone that is highly variable in all respects. None of the serpentines in this report are the fibrous or waxy varieties that are widely employed as polished building interiors, nor do they contain calcite veins (no cold HCl reaction), as described by Winkler (1994, p. 30). Rather, these building-face stone blocks are granular (0.1–0.4 mm diameter) and highly porous. A single quarry-fresh serpentine sample measured here had a bulk porosity of 12%. Composition of serpentine rocks from quarries in Nottingham, Pennsylvania (Table 1), 40 km from the quarried stone studied

TABLE 1. COMPOSITION OF SERPENTINE FROM WOOD'S AND OTHER QUARRIES NEAR NOTTINGHAM, PENNSYLVANIA

Mineral	Chemical Formula	Mean %	% Range
Serpentine	$Mg_6Si_4O_{10}(OH)_8$	83.1	16.3–96.7
Opaque minerals		5.8	0.2–9.4
Carbonates		1.9	0–24.6
Dolomite	$CaMg(CO_3)_2$		
Magnesite	$MgCO_3$		
Brucite	$Mg(OH)_2$	1.5	0–6.4
Olivine	$(Mg,Fe)_2SiO_4$	3.0	0–20.6
Chlorite	$(Mg,Fe)_5(Al,Fe)_2Si_3O_{10}(OH)_8$	1.9	0–75.7
Chromite	$FeCr_2O_4$	1.1	0–17.4
Late Crysotile	$Mg_3Si_2O_5(OH)_4$	0.4	0–3.3
Talc	$Mg_3Si_4O_{10}(OH)_2$	0.1	0–2.4
Others		1.1	0–25.8

Note: Data from McKague (1964). The number of thin sections analyzed was 40, and the number of points per thin section was 2500.

here, was determined from 40 thin sections (McKague, 1964). Most stone samples are made of serpentine, specifically lizardite (Smith and Barnes, 1998). Acid-susceptible secondary carbonate materials (dolomite and magnesite) on average make up just 2% of the stone, but percentages are highly variable (Smith and Barnes, 1998), ranging from 0 to 46% (McKague, 1964). In this study, many portions of serpentine blocks reacted slightly to a weak solution of warmed HCl, verifying the presence of at least some carbonate materials subject to acidic deposition.

History of Use

As early as the 1700s, serpentine was locally quarried as freestone for a few rock rubble structures in the West Chester, Pennsylvania, region (Ball, 1970). Taylor's quarry in West Chester, Pennsylvania, provided a pale, bluish-gray serpentine for today's perfectly preserved Chester County Historical Society building (erected 1848). Perhaps the durability of that stone encouraged further use of the inferior serpentine characterized here.

The serpentine described in this report is from Brinton's (then also variously known as "Birmingham," "Ingram's," "Thornbury," or "Westtown") quarry 5 km south of West Chester, Pennsylvania, which was used in buildings and other structures from 1870 to 1900. In contrast to the normal white and gray limestones and marbles of the Victorian Era, serpentine "suited the romantic architectural ideas of the late 1800's perfectly" (Ball, 1970). While mostly used for construction of wealthy homes, barns, spring houses, and university buildings near West Chester, the stone also enjoyed wide but scattered distribution in eastern Pennsylvania, Delaware, Chicago (Stone, 1932), and Hoboken, New Jersey (Julien, 1883). This author has identified more than 66 Brinton's quarry serpentine buildings in the West Chester region and 18 buildings farther afield (Table 2). Originally, there were more, but some buildings have been demolished.

Building stones in rural regions show that the Brinton-quarried ashlar blocks were surface-finished in many ways, presumably to the client's individual specifications. Most blocks have deeply etched impact tool marks or are so rough as to appear unworked. A few have near-smooth surfaces. Most are rectangular and vary in size from 10 to 60 cm across.

PHYSICAL DESCRIPTION OF SERPENTINE BLOCK WEATHERING

All serpentine blocks that retain their original impact tool marks (indicating no significant surface loss) currently display pale yellow hues (e.g., Munsell color 2.5Y 8/2.5), as opposed to the quarry-fresh, dark green colors (e.g. 10Y 5.5/2) that probably originally induced clients to select the stone for their building facings. Thus, all buildings in this study are visually "weathered," whether or not they have experienced surface recession.

TABLE 2. LATITUDE (NORTH) AND LONGITUDE (WEST) GPS COORDINATES FOR STRUCTURES MADE OF BRINTON'S QUARRY (39.9149, 75.5947) SERPENTINE BLOCKS

Houses, springhouses, garages, sheds, and walls:	(30.8998, 75.5700*), (39.9662, 75.6117), (39.9652, 75.6117*), (39.9644, 75.6121*), (39.9647, 75.61265*), (39.9303, 75.6009*), (39.9239, 75.5981), (39.9212, 75.6005), (39.9189, 75.60635), (39.9056, 75.6084), (39.9041, 75.6122), (39.9055, 75.5943), (39.9047, 75.5931), (39.9047, 75.5931), (39.9035, 75.5906), (39.9036, 75.5887), (39.9197, 75.5776), (39.9035, 75.5913*), (39.9025, 75.5901), (39.9293, 75.5561), (39.9158, 755860), (39.9126, 75.5931), (39.9126, 75.5930), (39.9171, 75.5947*), (39.9171, 75.5947), (39.9186, 75.5952), (39.9276, 75.5831), (39.1632, 75.5285), (39.9504, 75.2075*), (39.9554, 75.2050*), (39.9344, 75.1701*), (39.9736, 75.2025), (40.6802, 76.2019*).
Business buildings:	(39.9620, 75.6024), (39.9619, 75.6008), (39.6854, 75.7496*).
Barns:	(39.9121, 75.6005), (39.9158, 75.5860).
University buildings:	(39.9520, 75.5991*), (39.9514, 75.5986), (39.9528, 75.5998), (39.9513, 75.1944).
Churches:	(39.8998, 75.5700), (39.9566, 75.6023*), (39.3014, 75.6085*), (39.4508, 75.7152*), (39.7469, 75.5524*), (39.9354, 75.1755*).
Railroad station:	(39.9002, 75.6255).

*Indicates that the structure was measured and used in this report.

The yellow surface color of serpentine is probably due to alteration of iron-rich olivine and chromite to limonite. Surface colors tend to be more intense and reddish (e.g., 7.5YR 8/4) in regions where former air pollution was extreme. The yellow surface material often forms a thin (<2 mm), hard, protective surface. Studies on porous sandstone have shown that "case-hardened" layers are formed when dissolved ions from the stone interior migrate to the evaporative surface, where they are reprecipitated (e.g., Robinson and Williams, 1998). Hand lens views of spalled serpentine flakes support that observation. Yellowish outer surfaces are compact, but dark green underlying layers are more visibly porous than fresh quarry samples. X-ray diffraction suggests that the case-hardened surface is composed primarily of lizardite (some in a form approaching a gel) and primary crystalline chlorite.

A large number of blocks have lost portions of their yellow surfaces, either through sudden surface exfoliation of the case-hardened layer or by gradual granular weathering of surfaces that, for some reason, did not case-harden. Following loss of the yellow surface, green, porous, and presumably weakened serpentine is then exposed to the environment (Fig. 1), which produces the granular disintegration, flaking, cracking, and spalling that result in rapid surface recession and locally deep pitting (Fig. 2). A similar pattern of building stone breakdown occurs on some sandstones and granites (Amoroso and Fassina, 1983; Smith et al., 1994; Winkler, 1994; Robinson and Williams, 1996).

NON-QUANTITATIVE EVIDENCE OF SERPENTINE WEATHERING DUE TO ACIDIC DEPOSITION

Historical Anecdotes

Links between air pollution and serpentine deterioration were not immediately appreciated in the Philadelphia region. Twenty years after initial application of Brinton's serpentine in buildings, Merrill (1891, p. 373) noted, "Serpentine is non-absorptive and not affected by gaseous atmospheres, hence it is durable if free from bad veins or joints. Pennsylvania serpentine may fade or turn whitish, but, so far as observed, does not disintegrate."

By the early 1900s, however, the Topographic and Geologic Survey Commission of Pennsylvania (Anonymous, 1913) stated, "Serpentine being chiefly of a magnesian nature is easily affected by certain sorts of acid vapors in the air, such as sulphurous ones. These sulphurous vapors are not especially noticeable in the rural districts but are pronounced in cities where smoke and vapors from chemical and other manufacturing plants are very common."

In the 1930s, a time of maximum SO_2 emissions in the northeastern United States (Lipfert, 1987), Heathcote (1932, p. 8) wrote, "Serpentine houses in Chester County are in many cases more than 100 years old, and are still in a fresh and sound condition, due no doubt, to the air which is free from smoke and fumes of all sorts. In Philadelphia, where serpentine has been used for buildings such as some of those at the University of Pennsylvania, vapors in the air, particularly sulphurous ones from manufacturing plants, oil refin-

Figure 1. A serpentine building in Pottsville, Pennsylvania, displays a pale (yellowish), case-hardened original surface with tool marks, as well as a dark (green), receded serpentine surface near a window frame. Serpentine grains and flakes resting on window sills show that the weathering is still occurring.

Figure 2. Granular weathering and exfoliation of small flakes have created a 61-mm-deep pit in a serpentine block on Grace Methodist Church, Wilmington, Delaware.

eries and railroads nearby, seriously affect the composition of the serpentine, causing it to crumble."

Buildings "Missing in Action"

Probably the most expansive (and expensive?) attempt to retain the historically valued serpentine tradition is evident at the original core of the University of Pennsylvania. For example, Logan Hall, formerly serpentine, has been completely rebuilt with green cement blocks. The north, east, and south walls of College Hall are now a mixture of green block–replacement strategies. Only the west wall of College Hall retains a few (highly-weathered) serpentine blocks.

Some serpentine buildings in downtown Philadelphia, listed in earlier inventories, are now absent. Stone (1932) mentions that there are "many" serpentine buildings and residences in Philadelphia. This author found only seven after reconnoitering almost all streets in the city. Other listed (Merrill, 1891; Stone, 1932) important buildings that no longer exist include Memorial Baptist Church, Holy Communion Church, and Academy of Natural Sciences.

Buildings in regions of moderate former air pollution are also missing, including Second Presbyterian Church in Pottsville, Pennsylvania, the serpentine of which Stone (1932, p. 262) described as "not a good stone in the Pottsville atmosphere. The dressed stone is spalling, and the rubble is a horrid yellowish green." A more notable example removed in 1956 is "Old Main" gymnasium (1876) on the campus of West Chester State University, Pennsylvania, said to be the "the largest serpentine building in the world" (Ball, 1970). It was demolished despite public outcry because its weathered serpentine exterior was thought to indicate loss of structural integrity underneath. Such did not prove to be the case, so nearby Ruby Jones and Recitation Halls were spared from destruction and remain serviceable today.

Continuing Surface Loss

Serpentine grains and flakes are still falling, as seen on ground and window sills (Fig. 1) in formerly polluted locations (Philadelphia, Pottsville) but not in rural locations. Such surface loss could represent the "memory effect," wherein present weathering processes work more rapidly in materials previously made porous and weak during extreme stone disintegration events of the past.

SERPENTINE BLOCK SURFACE RECESSION MEASUREMENT METHODOLOGIES

To assess the extent and severity of material loss on serpentine building stones, a survey was conducted in which the maximum depth of cavities on block surfaces was measured and the areal extent of surface recession was observed on a total of 6000 individual blocks. Mean ($n = 10$) maximum pit depth (Fig. 2) per building side was measured with a depth probe (1×3 mm rectangular tip) inserted into pits 1.5–2.5 m above the ground.

The initial block surface against which to measure pit depth was established in several ways. For building sides that are moderately weathered (most), many block surfaces still retain their original tool marks (Fig. 1). On those walls, a visible (eye at right angle to the wall) projection of high points of the original tooled surface becomes the initial reference plane from which mean measured tool mark pit depth is subtracted. For more weathered buildings that lack tool mark surfaces, an original, durable reference surface could almost always be found against which to measure pit depth (e.g., window adornments; Fig. 2). Only twelve buildings (16 sides) had both direct access for pit depth measurements and clear initial block surfaces.

A second index of serpentine surface loss needs no direct building access. Areal percentage lacking original tool marks was estimated for each building block from a ×8 binocular view. Assessments were made from many blocks across an entire building side at 2 m height intervals (judged by reference to the author's 2 m height or to tape placed at 2 m on the building). Every block ~0.3 m above and below each 2 m horizontal datum was assessed. The mean areal percentage of building blocks lacking tool marks was computed for each 2 m vertical interval on each building side. Tool mark pit depths vary greatly from building to building, so an original surface roughness of ~10 mm was arbitrarily selected as the maximum for inclusion of a building into the study. This restriction, plus the fact that some buildings were originally built from blocks without impact tool marks, included 22 buildings (38 building sides, 6000 blocks) in the study.

These two field measurements of serpentine block surface loss are only surrogates for volumetric loss and are semiquantitative at best. The rate at which serpentine blocks disintegrate varies widely, even between adjacent blocks on a wall, due both to extreme textural and compositional stone nonhomogeneities (Table 1) and to complicated building geometry. Second, only building blocks larger than 20 cm across were measured here, but block sizes vary greatly above that minimum, so measurements made "per block" are statistically biased between buildings with larger or smaller blocks. Third, large buildings have more blocks to measure than do small houses, leading to sample size comparison problems.

NON-POLLUTION SERPENTINE WEATHERING PROCESSES

Rainwater Flow

Serpentine deterioration on each building is visibly greater where the flow of rainwater is enhanced, such as around windows, chimneys, and other vertical protuberances that have channeled runoff. Stones are often in poor condition around drainpipes, either because gutters fell into disrepair and emptied excess rainwater into the serpentine or because excess condensation on the metal enters the surrounding stone (Winkler, 1994).

Semiquantitative measurements of weathering by enhanced water flow include the south side of Grace Methodist Church in

Wilmington, Delaware, where a section with overhanging eaves has only 54% of its block area intact (i.e., lacking tool marks), compared to a non-eaved section with 97% of the block area deteriorated. Similarly, an eaved wall on the south face of Recitation Hall, West Chester, Pennsylvania, has lost 53% of stone tool marks, but the non-eaved part 77%.

Groundwater Wicking ("Rising Damp")

As is the case with many porous stone types (e.g., Amoroso and Fassina, 1983; Winkler, 1994; Gauri and Bandyopadhyay, 1999), serpentine buildings are visibly more weathered near the ground than at a height of 1–2 m. Measurements of 12 serpentine building sides that extend into the soil (Fig. 3) show that near-ground surface loss is not only greater near the ground than 2 m higher, but this effect is generally more pronounced on buildings that are more weathered overall.

Building Side Orientation and Serpentine Weathering

The only three buildings (Grace Church, Wilmington, Delaware; Recitation Hall, West Chester, Pennsylvania; and St. Anne's Episcopal Church, Middletown, Delaware) that are faced with serpentine on all four sides display identical aspect-related weathering patterns. The mean areal percentage of serpentine blocks lacking tool marks, pooled for the three buildings, demonstrates that south-facing (75%) and east-facing (60%) sides are more weathered than are west-facing (35%) and north-facing (29%) sides.

The reasons for these large aspect differences in surface recession are unclear. Frost weathering should affect the north wall more than the south, and insolation weathering should affect the west sides more than the east, but this does not explain the above findings. Theoretically, wind direction and velocity should augment driving rain (Sherwood, 1990) and dry deposition rates of acidic materials (Haber et al., 1988), but predominant winds are from the wrong direction in the Philadelphia region to invoke this explanation. Prevailing wind directions (1930–1996 data from National Climatic Data Center [2005]) are from the NNW in winter and spring and from the SW in the summer and fall, so the west-facing walls should be the most weathered, but they are in relatively good condition. In this case, it is likely that south-facing and east-facing stone blocks experience earlier and more frequent surface drying by insolation on a daily basis. Surface drying would promote salt crystallization at, or near, the surface and also promote freeze-thaw cycles during the winter months, thereby enhancing material loss in the long-term.

Interactions between Serpentine and Adjacent Carbonate-Rich Materials

Winkler (1982) pointed out that calcium carbonate is leached from mortar joints by rainwater and can be reprecipitated on nearby porous stone. Such a process may render

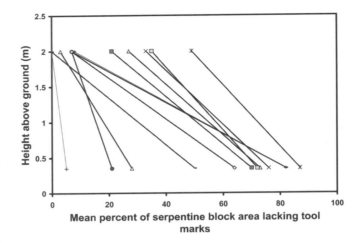

Figure 3. Greater weathering of serpentine blocks due to groundwater wicking near the ground than at ~2 m height is demonstrated by mean areal percent loss of tool marks for blocks at 12 different building sides.

porous serpentine, which generally contains few acid-reactive minerals, more susceptible to gypsum growth and consequent mechanical attack. Although not yet measured, serpentine surfaces visually do appear more receded near lime mortar than in block centers, as would be expected from other research (Brown and Clifton, 1988).

Two marble sills (4 m and 8 m above the ground) encircle Recitation Hall, West Chester, Pennsylvania (Fig. 4). Serpentine blocks directly underneath them are 2–3 times more damaged than those above, on all four sides of the building (Fig. 5), strongly suggesting that dissolved lime from the sills enters the serpentine and augments its susceptibility to weathering.

Figure 4. Surface loss from serpentine blocks is visibly greater directly under two marble sills than it is elsewhere on Recitation Hall, West Chester State University, West Chester, Pennsylvania.

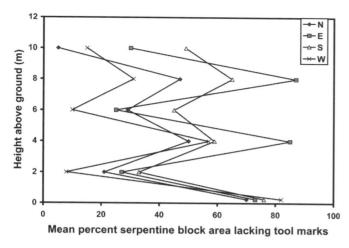

Figure 5. Surface loss (tool marks absent) is significantly greater for blocks under two marble sills (4 and 8 m above the ground) than it is above or well below the sills at Recitation Hall, West Chester, Pennsylvania. Apparent enhancement of serpentine weathering above the sills is an artifact of the vertical sampling interval. Groundwater wicking accounts for the greater weathering near ground level.

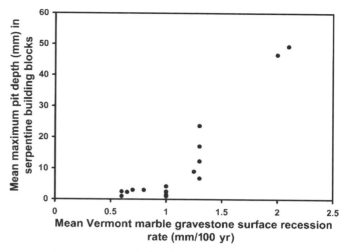

Figure 6. In the greater Philadelphia region, mean ($n = 10$) maximum pit depths in serpentine blocks (each dot represents one wall) generally increase with amount of historic acidic deposition (indicated by surface recession rates on nearby acid-sensitive Vermont marble slab tombstones).

INFLUENCE OF ACIDIC DEPOSITION ON SERPENTINE SURFACE LOSS

Indices of serpentine weathering (mean maximum pit depth, mean areal percent of blocks lacking tool marks) were computed for each available building side, and these values are graphically correlated with the acidic deposition surrogate: Vermont marble tombstone mean surface recession rates (Figs. 6 and 7) for the entire Philadelphia region. If a serpentine building is within a kilometer of a suitable cemetery, that local marble surface recession rate is applied. Buildings farther from cemeteries with marble stones are assigned an interpolated (SURFER program–inverse squared distance selection) marble weathering rate based upon all cemeteries within 10 km.

A major assumption is that a computed mean based on a large number of measured blocks (excluding those in the groundwater wicking zone) on each building side can overcome the many confounding variables of aspect, building geometry, and nonhomogeneity of stone composition and dressing. However, highly variable sample sizes ($n = 10–649$) from one building side to another statistically invalidate numerical regression and correlation analyses. The visual correlations (Figs. 6 and 7) remain semiquantitative only.

Weathering pits in serpentine blocks (Fig. 6) are generally small and shallow in regions where marble tombstones received little acidic deposition (as indicated by marble weathering). Weathering pits are much deeper and more numerous on building sides that received large doses of acidic gases and/or acidic precipitation, such as downtown Philadelphia and Wilmington, Delaware (Fig. 2).

The correlation between serpentine block surface loss (lack of original tool marks) and pseudo-SO_2 input is visibly apparent

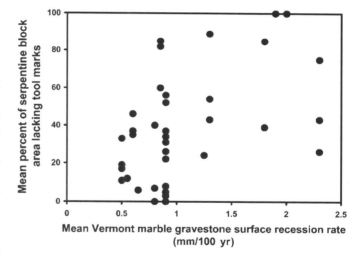

Figure 7. In the greater Philadelphia region, rural serpentine building blocks, which have received low acidic deposition doses (as indicated by low marble tombstone surface recession rates) have lost little of their original surfaces (mean percent block area lacking tool marks). Where acidic deposition was greater, serpentine blocks have lost more of their original surfaces.

but weak (Fig. 7). The largest, positive exception to the positive air pollution–serpentine recession trend is a church in Smyrna, Delaware (0.85 mm/100 yr of nearby marble tombstone surface recession), a town with no known historical coal-fired industries. Likewise, but as a negative residual, several serpentine homes near the very weathered serpentine buildings of the University of Pennsylvania (1.8 mm/100 yr of marble tombstone surface recession) are in relatively good condition, perhaps because they are in a wealthy neighborhood with solid long-term repair histories.

The most weathered serpentine building still standing is the 19th and Titan Street Baptist Church (constructed 1874) in southwestern Philadelphia (Fig. 8). Its thick, green-colored cement exterior, meant as a repair (date uncertain, but thought to be in the 1970s), has now become exfoliated over approximately half the building surface area. The original vertical plane surface of the smoothly cut serpentine block surface is the same as that of the cement exterior, as revealed from a straight horizontal line of serpentine adjoining the newer concrete sidewalk. The current position of the exposed serpentine surface shows that the surface loss was already extreme (~6 cm, or 60 mm/100 yr) before the cement covering was added.

Figure 8. Strongly weathered serpentine blocks on the 19th Street Baptist Church in Philadelphia were covered at some time in the past with repair cement that has since exfoliated. The outer face of the cement is at the same vertical plane as the original serpentine block face and so demonstrates the extreme amount of serpentine surface recession that occurred before the repair cement was applied.

DISCUSSION AND CONCLUSIONS

More than a century ago, a few wealthy clients in the Philadelphia region selected dark green serpentine stone from a single quarry to face their buildings. Some of these buildings have been replaced, and others are in danger of removal because of massive block deterioration. As is the case with many rare stone cultural objects, preservation demands knowledge of deterioration mechanisms. This report uses geographic variations in serpentine weathering to test the hypothesis that acidic deposition augments the stone damage.

It is a small wonder that buildings have seldom been used to spatially test the concept that air pollution, or indeed any other process, has been responsible for stone deterioration. Stone weathering is by nature multivariate. Buildings introduce a complexity of form far beyond that of slab tombstones, which have previously demonstrated that the geographic distribution of deterioration is closely related to the regional distribution of acidic deposition. The Philadelphia region, with its relatively well-known cumulative SO_2 input, acts as an outdoor analogue to a chemist's environmental chamber, but it proved difficult in this study to hold constant all building variables except for air pollution. Both indoor laboratory and outdoor geographic approaches encounter the difficulty of accounting for initial heterogeneity in the properties of a named, classified stone (in this case, "serpentine"). High block-to-block weathering variability on a single building side is not unique to serpentine; it has also been observed on buildings made of sandstone (Robinson and Williams, 1996; Turkington and Smith, 2004) and granite (Urquhart et al., 1996)

A further complication in the "real world laboratory" is that historic stone structures cannot escape temporal, geometric, and aspect variations in the delivery rate of destructive gases and liquids to individual stones and buildings. Buildings have unknown repair histories for eaves, gutters, mortar, and blocks, and sides differ in solar and rain input. It appears that the number of variables on a building far exceeds the number of buildings available to adequately standardize the many non-acidic deposition variables.

Even in rural buildings, rainwater can enter porous serpentine stones from above, and soil water rises from below. Wherever water intrudes porous, structurally weak stone, dissolution, hydration, and oxidation of minerals, plus mechanical breakdown due to expansive formation of ice and salts (crystal growth, thermal expansion, hydration) are likely. It is therefore surprising that the field-measured data (especially mean maximum pit depth) collected here generally support anecdotal accounts and circumstantial evidence that "sulphurous" air pollution or acidic precipitation has augmented serpentine weathering in urban/industrial settings.

Whereas the amount of surface loss to serpentine buildings in central Philadelphia is greater than this author has observed for any other stone type in North America, the way in which acidic deposition causes the damage remains uncertain. It is tempting to state that SO_2, catalyzed by NO_x, reacts with the carbonates in the serpentine to form gypsum, which causes physical weathering, but why should the damage be greater in a stone with <25% (usually much less) carbonates than in marble or limestones with nearly 100% calcite? Further, rain-protected sites on marble, limestone, and some sandstone types produce the notoriously destructive, gypsum-based "black crust" when exposed to high levels of SO_2 and carbonaceous fly ash (e.g., Amoroso and Fassina, 1983; Gauri and Bandyopadhyay, 1999), but this does not occur under eaves in serpentine buildings. Perhaps acidic wet deposition is responsible for much of the serpentine deterioration, but again, the mechanisms are uncertain because there is no visible evidence of dissolution loss of serpentine in either city or countryside. The processes by which groundwater wicking produces more serpentine damage in city than countryside remain unclear and may have been enhanced by the seasonal input of deicing salts in the urban environment.

This empirical, geographic study of serpentine weathering has produced more mystery than revelation about the processes of stone weathering, but it pinpoints locations where chemists might best collect and analyze reaction products that indicate the causes of deterioration. The architecturally elegant and rare serpentine buildings deserve further attention so that they can be more knowledgeably protected.

ACKNOWLEDGMENTS

Thanks to Peter Leavens, Geology Department, University of Delaware, for analyzing serpentine samples. Jane Dorcester, University of Pennsylvania, characterized serpentine use in architecture. Several unsung heroes (anonymous reviewers) aided in text revision.

REFERENCES CITED

Amoroso, G., and Fassina, V., 1983, Stone decay and conservation: New York, Elsevier, 453 p.

Anonymous, 1913, Topographic and Geologic Survey Commission of Pennsylvania, Report 9.

Ball, B.M., 1970, Chester County and its day: Chester County Day Committee of the Women's Auxiliary, The Chester County Hospital, 76 p.

Baer, N.M., and Berman, S.M., 1983, Marble tombstones in national cemeteries as indicators of stone damage: general methods: Preprint 83-5-7: Atlanta, Annual Meeting of the Air Pollution Control Association, 24 p.

Brown, P.W., and Clifton, J.R., 1988, Mechanisms of deterioration in cement-based materials and in lime mortar: Durability of Building Materials, v. 5, p. 409–420.

Charola, A.E., 1988, Chemical-physical factors in stone deterioration: Durability of Building Materials, v. 5, p. 309–316.

Charola, A.E., 1998, Review of the literature on the topic of acidic deposition on stone: U.S. Department of Interior, National Center for Preservation Technology and Training Publication 1998-09, 85 p.

City of Philadelphia, 1983, Emissions inventory and air quality data report to the Air Pollution Control Board: Philadelphia, Air Management Service, Department of Public Health, 32 p.

Dann, K.T., 1988, Traces on the Appalachians: a natural history of serpentine in eastern North America: New Brunswick, Rutgers University Press, 159 p.

Feddema, J.J., and Meierding, T.C., 1987, Marble weathering and air pollution in Philadelphia: Atmospheric Environment, v. 21, p. 143–157, doi: 10.1016/0004-6981(87)90279-4.

Gauri, K.L., and Bandyopadhyay, J.K., 1999, Carbonate stone: New York, John Wiley & Sons, 284 p.

Haber, J., Haber, H., Kozlowski, R., Magiera, J., and Pluska, I., 1988, Air pollution and decay of architectural monuments in the city of Cracow: Durability of Building Materials, v. 5, p. 499–547.

Heathcote, C.W., 1932, A history of Chester County, Pennsylvania: Harrisburg, Natural History Association, 132 p.

Johansson, L.G., Lindqvist, O., and Manio, R.R., 1988, Corrosion of calcareous stones in humid air containing SO_2 and NO_2: Durability of Building Materials, v. 5, p. 439–449.

Julien, A.A., 1883, The decay of the building stones of New York City: Proceedings of the American Association for the Advancement of Science, v. 28, p. 372–383.

Küpper, M., 1975, Récherches en Belgique sur l'alteration des pierres calcaires exposeés a l'air libre: Lithoclastia, v. 2, p. 9–18.

Lipfert, F.W., 1987, Estimates of historic urban air quality trends and precipitation acidity in selected U.S. cities (1880–1980): Brookhaven National Laboratory, Report 39845, 14 p.

Mather, J.R., 1968, Meteorology and air pollution in the Delaware Valley: Publications in Climatology, v. 21, 136 p.

McKague, H.L., 1964, The geology, mineralogy, petrology, and geochemistry of the state line serpentinite and associated chromite deposits [Ph.D. Dissertation]: State College, The Pennsylvania State University, 164 p.

Meierding, T.C., 1993, Marble tombstone weathering and air pollution in North America: Annals of the Association of American Geographers, v. 83, p. 568–588.

Meierding, T.C., 2000, Philadelphia's effect on precipitation acidity from marble gravestone dissolution rates: The Pennsylvania Geographer, v. 38, p. 42–56.

Meierding, T.C., 2004. Arkose 'Brownstone' tombstone weathering in the Northeastern USA: *in* Smith, B.J. and Turkington, A.V., eds., Stone decay: its causes and controls: London, Donhead, p. 167–197.

Merrill, G.P., 1891, Stones for building and decoration: New York, John Wiley, 453 p.

National Climate Data Center, 2005, http://www5.ncdc.noaa.gov/documentlibrary/pdf/wind1996.pdf.

Price, C.A., 1996, Stone conservation: an overview of current research: Santa Monica, California, The Getty Conservation Institute, 73 p.

Robinson, D.A., and Williams, R.B.G., 1996, An analysis of the weathering of Wealden sandstone churches, *in* Smith, B.J. and Warke, P.A., eds., Processes of urban stone decay: London, Donhead, p. 133–149.

Robinson, D.A., and Williams, R.B.G., 1998, The weathering of Hastings Beds sandstone gravestones in south east England, *in* Jones, M.S. and Wakefield, R.D., eds., Aspects of stone weathering, decay and conservation: London, Imperial College Press, p. 1–15.

Schreiber, K.V., and Meierding, T.C., 1999, Spatial patterns and causes of marble tombstone weathering in western Pennsylvania: Physical Geography, v. 20, p. 173–188.

Sherwood, S., 1990, Processes of deposition to structures: Washington, D.C., National Acid Precipitation Program, State of Science and Technology, Report 21, 145 p.

Smith, J., Magee, R.W., and Whalley, W.B., 1994, Breakdown patterns of quartz sandstone in a polluted urban environment: Belfast, Northern Ireland, *in* Robinson, D.A. and Williams, R.B.G., eds., Rock weathering and landform evolution: Chichester, John Wiley & Sons, p. 131–150.

Smith, R.C., and Barnes, J.H., 1998, Geology of Nottingham Park: Pennsylvania Department of Conservation and Natural Resources Open File Report 98-12, 41 p.

Stone, R.W., 1932, Building stones of Pennsylvania: Pennsylvania Geological Survey Bulletin M15, 316 p.

Turkington, A.V., and Smith, B.J., 2004, Interpreting spatial complexity of decay features on a sandstone wall: St. Matthew's Church, Belfast, *in* Smith, B.J. and Turkington, A.V., eds., Stone decay: its causes and controls: London, Donhead, p. 149–166.

Urquhart, D.C.M., Young, M.E., MacDonald, J., Jones, M.S., and Nicholson, K.A., 1996, Aberdeen granite buildings: a study of soiling and decay, *in* Smith, B.J. and Warke, P.A., eds., Processes of urban stone decay: London, Donhead, p. 66–77.

Winkler, E.M., 1982, Decay of stone monuments and buildings: the role of acid rain: Technology and Conservation, v. 7, p. 32–41.

Winkler, E.M., 1994, Stone in architecture: properties, durability: Berlin, Springer-Verlag, 313 p.

MANUSCRIPT ACCEPTED BY THE SOCIETY 19 JANUARY 2005

Geological Society of America
Special Paper 390
2005

Surface-recession weathering of marble tombstones: New field data and constraints

Sheila M. Roberts

Department of Environmental Sciences, The University of Montana–Western, Dillon, Montana 59725, USA

ABSTRACT

Vertical marble-slab tombstones offer unique advantages over natural stone exposures for measuring historical carbonate weathering rates. One of the most commonly measured properties is surface recession rate, defined in this study as the difference in thickness between the relatively unweathered base and the weathered top of the tombstone.

Use of the "base-minus-top" method makes several assumptions, which are revisited in the light of this research:

1. Tombstone thickness from top to base is originally very uniform. Data show that up to 1 mm (but less than 2 mm) original difference in thickness may be common.

2. The tombstone base remains sufficiently unweathered to provide an accurate estimate of the original thickness. Different locations have different maximum ages for which this assumption is valid.

3. Weathering is even across the top of the tombstone. Although thickness across the top of an individual weathered tombstone varies, three consistently placed measurements seem to provide a reproducible average thickness.

4. Surface recession at the tombstone top is measurable and acceptably precise. The total thickness loss must be significantly greater than the original thickness difference and the experimental error.

5. Weathering is dominated by loss of thickness of the marble slab. This assumption is generally correct over long periods of time. However, the data reported here indicate early-stage expansion of some exposed surfaces.

6. Surface recession is linear over time. This study documents nonlinear surface recession of tombstone surfaces.

Keywords: carbonate weathering, dissolution, marble tombstones, surface recession rate, base-minus-top method, standard error, cemetery, tombstone thickness, nonlinear weathering, weathering rates.

INTRODUCTION

Quantifying recent weathering rates of carbonate rocks exposed at Earth's surface has important implications for our ability to respond to the destruction of public buildings and monuments by weathering. These data also inform our understanding of the dynamics of the carbon cycle and associated climate change, the formation of soils, and the evolution of landscapes. In addition, comparisons of historical weathering rates with those in the more distant past may assist scientists in quantifying the influence of human activities on these processes.

Roberts, S.M., 2005, Surface-recession weathering of marble tombstones: New field data and constraints, *in* Turkington, A.V., ed., Stone decay in the architectural environment: Geological Society of America Special Paper 390, p. 27–37, doi: 10.1130/2005.2390(04). For permission to copy, contact editing@geosociety.org. ©2005 Geological Society of America.

The potential advantages of using marble tombstones to measure weathering rates, listed elsewhere (see reviews by Livingston and Baer, 1990; Pope et al., 2002), include (1) material is abundant and readily available; (2) the date of exposure of a "fresh" surface is (usually) easily determined; (3) identical or similar lithologies are emplaced in diverse settings, which makes environmental comparisons feasible; and (4) several aspects of tombstone weathering record measurable change within 100 yr or less of exposure (surface recession rate, inscription depth and legibility, corner rounding, relief of lead lettering or inclusions, exfoliation area, etc.). This paper summarizes and critiques assumptions inherent in one of the most widely applicable and easily quantifiable methods for measuring the weathering rates of marble–surface recession rate of vertical marble-slab tombstones.

The basis of this critique is new field data collected in southwestern Montana in the western United States. Tombstones were measured at seven cemeteries serving four towns (Fig. 1) with broadly similar climates (Table 1). Average annual precipitation at three of the sites (Butte, Dillon, and Helena) is very close (27–32 cm), while Bozeman has substantially higher precipitation (47 cm). Average maximum temperatures at the four locations are all within 2.6 °C and average minimum temperatures are within 2.9 °C. The cultural setting of the towns varies from rural to urban-industrial.

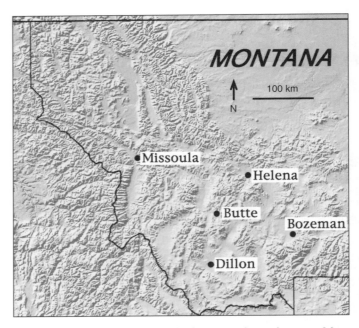

Figure 1. Location map showing the four towns in southwestern Montana where tombstones were measured (Butte, Bozeman, Helena, and Dillon). Missoula is included on the map for reference.

THE "BASE-MINUS-TOP" METHOD OF DETERMINING SURFACE RECESSION

Several types of measurements of surface recession of tombstones are described in the literature (e.g., Dragovich 1987, 1991), based on the inference of some weathering process that results in loss of surface material. All methods require a measurable property that is related to weathering and some frame of reference back to the original unweathered state.

The different weathering rates of tops and bases of vertical tombstones is easily observed in most cemeteries by comparing visible differences in color, surface polish, cracking, corner angularity, etc. (Fig. 2), but these factors are difficult to quantify. The discussion in this paper is confined to an established method that

compares the top thickness (maximum weathering loss) to base thickness (minimum weathering loss) of marble-slab tombstones to determine surface recession (as described by Meierding, 1981; Baer and Berman, 1983; Feddema and Meierding, 1987).

Theoretical Basis

The simplest interpretation of the base-minus-top recession data is that the top of the tombstone is more weathered than the base because the top is more directly exposed to wet acidic precipitation (rain or snow). The acidity of precipitation is buffered by flow across the marble top and sides of the tombstone before it reaches the base. As a result, the base is protected from weath-

TABLE 1. COMPARISON OF SOME ENVIRONMENTAL FACTORS
AT CEMETARY LOCATIONS IN SOUTHWEST MONTANA

Location	Average annual precipitation. (cm)	Average T (°C)		Approximate altitude (m)	Rural	Urban	Industrial	Time of record
		maximum	minimum					
Bozeman	46.51	12.7	−0.6	1500	X	X		1950–2004
Butte	32.4	11.8	−2.8	1700		X	X	1894–2004
Dillon	29.03	14.4	−1.7	1550	X			1895–2004
Helena	27.33	12.9	0.1	1250		X	X	1893–2004

Note: Climate data from Western Regional Climate Center, 2004, Montana Climate Summaries: Desert Research Institute (www.wrcc.dri.edu/summary/climsmmt.html).

Figure 2. A 1902 tombstone in the Dillon cemetery, with close-up views that show differential weathering of the top and base. The top of the tombstone is roughened and is developing shallow cracks at the edges. Corners are beginning to be rounded and there is discoloration. The base retains its original white color and surface polish.

ering. In addition, the role of rainfall in removing the reaction products that interfere with further weathering (Feddema and Meierding, 1987) will be more effective at the top surface, where rain hits the stone first. For any individual tombstone, it is theoretically possible (and relatively easy) to determine maximum surface recession by measuring the presumed original thickness (as defined by the current base thickness) and comparing it to the weathered, recessed thickness at the top.

Some experimental work corroborates this assumption. For example, Reddy (1988) compared the composition of incident rainfall to the composition of runoff from exposed marble and limestone surfaces. He found that surface recession (as calculated from increased Ca^{2+} in the runoff) was strongly correlated with the amount of rainfall and its pH, and not strongly correlated with several other factors, including lithology.

Rock weathering is a combination of physical, chemical, and biological processes that operate at different rates, over different time spans, in different environments. An underlying assumption of the method, not tested here, is that in-place dissolution of crystals is the dominant process causing surface recession of

marble tombstones in this area during the first 100 years or so of exposure. This assumption is part of the theoretical basis for predicting that the tops of the tombstones will lose thickness much more rapidly than the bases. The assumption of the dominant effect of solution goes back as far as Geikie (1880), although he noted internal disintegration, exfoliation, fracture, and loss of whole crystals. In the early stage of weathering, if whole-crystal loss and other processes create a very small-scale uneven surface, the measurements as described here will not "see" those losses because calipers cannot account for missing material below the crystals that protrude from the surface.

The base-minus-top method will not be a reliable approach to measuring surface recession anywhere weathering is dominated by a process that operates either unevenly or randomly across the entire tombstone face or selectively at the tombstone base instead of the top. Direct interaction with atmospheric gases, especially SO_2, may be one of those processes (e.g., Gauri et al., 1983, 1989; Feddema and Meierding, 1987; Skoulikidis and Charalambous, 1981), although other factors such as air circulation patterns may still leave the base relatively unweathered

(Feddema and Meierding, 1987). Field observations also suggest that in some environments biogenic weathering effects related to moss and lichen growth, which do not fit the model of top preference, may be very significant (e.g., Wakefield and Jones, 1998). Irrigation of cemeteries by sprinklers may throw water on any level of the tombstones and change the weathering pattern. Microscale loosening of crystals related to thermal expansion and contraction, salt crystallization and hydration, or frost heaving (e.g., VanGemert et al., 1988; Winkler, 1987; Warke, 2002) would not necessarily operate preferentially on the tops of the tombstones in all environments.

Another variable, which applies to tombstones in general, is that some freshly dressed tombstones may be already weakened by microcracks or microscopic loosening of crystals resulting from the transition from metamorphic conditions to surface conditions (Pope et al., 2002). This would not be noted in the field, but could significantly alter weathering rate and spatial distribution.

Method

The base-minus-top method of measuring surface recession used for this study is simple and inexpensive. Five measurements are collected at each vertical marble-slab tombstone: two at the base (one on either side, as near as possible to the ground) and three across the top (at the right and left edges and in the middle) (Fig. 3). This combination of measurements produces a comparison of the average top measurement with the average base measurement, which is interpreted as the maximum surface recession for that stone. Tombstones were initially (1999) measured with dial calipers accurate to ±0.001 inch and later (post 1999) with digital calipers accurate to ±0.02 mm. The earlier measurements were converted to millimeters.

Feddema and Meierding (1987) used a modified approach, in which their upper measurement was taken at 50 cm from the base of the stone. This measurement method assumes that (1) direct interaction with atmospheric SO_2 and water across the face of the tombstone is the dominant weathering process, and (2) microclimatic variables that control delivery of pollutants to the stone face are constant above ~30 cm. In the present study, the top of the stone was measured instead because (1) areas of stones protected from water flow by overhanging marble retained polish and a fresh appearance, suggesting that flow across the stone was more important than direct atmospheric interaction; (2) the top was usually visibly more weathered than the sides, and the intent was to estimate maximum weathering; and (3) if weathering diminishes down the stone, the tombstone top provides a more consistent measurement for comparing maximum weathering of stones of different heights than a measurement 50 cm from the base.

Although measurements in some cemeteries are not yet complete, the intent of this survey is to measure virtually all tombstones that fit the following criteria in every cemetery visited:

1. White to light gray, relatively pure calcite or dolomite marble (avoids weathering bias due to mineralogical variation and saves time).

 Measurement locations

Figure 3. Locations of the five measurements required for each tombstone (arrows on upper photograph), shown on a 1925 tombstone from Helena. Lower photograph demonstrates placement of calipers during a base measurement (in this case, below chipping caused by lawn mowers).

2. Slab shape (eliminates intentional manufacturing differences between top and base thickness and alterations of water flow related to tombstone geometry). Tops could be cut flat, rounded, or in somewhat more ornate designs, as long as the original slab thickness was not compromised.

3. Fine to coarse crystals (most stones were of fine to medium crystal size). Although crystal size may be significant, this study did not control for that factor.

4. Massive texture (avoids weathering bias due to crystal orientation or layering).

5. Vertical placement of the stone.

The decision to measure all tombstones that fit the criteria avoids problems of sampling bias. In most cemeteries visited, measuring all tombstones is not an overwhelming burden. Marble tombstones are common at veterans' graves but become progressively less common through time at other graves, replaced by granite or other lithologies.

The most problematical aspect of the fieldwork is deciding when a marble-slab tombstone should not be included in the data set. Criteria used for elimination include:

1. Lithologic inconsistencies (too many mineral impurities, extreme color mottling, foliation, other structural or lithologic heterogeneity).

2. Date or location inconsistencies (stone emplacement appeared to be significantly different than the date of death or there was evidence that the stone may have been relocated).

3. Obvious restoration activity (gluing, cementing, replacement of broken corners with new material, etc.).

4. Unusual micro-environmental conditions (stones that were much more lichen-covered than others in the cemetery, position under trees that would reduce contact with direct precipitation, etc.).

5. Non-vertical position (strongly tilted to horizontal). In one cemetery, vandals had broken the tops off many stones, and they were left tilted against the remaining base or lying on the ground).

6. Other factors. For example, in a few cases, lawn mowers had chipped the bases of stones so thoroughly that they could not be measured. Stones positioned for a long time near sprinkler heads were covered with mineral deposits. Dirt or vegetation buildup around tombstone bases occasionally prevented correct placement of the calipers.

CRITIQUE OF SOME ASSUMPTIONS OF THE "BASE-MINUS-TOP" METHOD OF MEASURING SURFACE RECESSION

Pope et al. (2002) and Livingston and Baer (1990) provided useful summaries of all the published methods for measuring surface recession of cultural stone and some of the assumptions for each. This paper expands on their discussion of the surface-recession method and focuses on the assumptions of the "base-minus-top" method.

1. Tombstone Thickness from Top to Base is Originally Very Uniform

This assumption can be tested only if it may be assumed that construction of recently emplaced marble tombstones is analogous with tombstone construction in the past. Measurements of

27 recently emplaced vertical marble tombstones show that original base versus top thickness differences may be significant. The new tombstones (1990–2001) had variations in original thickness differences clustering around ±0.50 mm (Fig. 4), but ranging up to +1.63/−0.72 mm. The standard deviation was 0.40 mm. Differences up to ±1 mm are probably fairly common in modern tombstone construction.

All but one of these new tombstones were from veterans' graves. According to Dennis Gerdovich (National Cemetery Administration, Washington, D.C., 1999, personal commun.), the U.S. government contracts the manufacturing of veteran's tombstones, specifying a thickness tolerance of 1/8 inch (3.2 mm). Livingston and Baer (1990) report a Veterans Administration allowance for a 4-inch (8.2 cm) tombstone of +1/4 inch (6.4 mm) and −1/8 in (3.2 mm). Apparently (and luckily) the actual thickness difference across these stones may be typically better controlled than the specifications require.

Part of the original thickness problem has been addressed by another assumption, utilized by this and other studies, that tombstones were probably engraved randomly with respect to original thickness differences up or down. That assumption is useful because the original differences (±) would then cancel each other out in calculations of mean values for large numbers of tombstones. More data from new tombstones may eventually show if there is a statistical bias toward thick end up or down.

Right-left differences in original stone thickness are also of interest in estimating overall manufacturing consistency. Left and right sides of 174 tombstone bases (of presumed original thickness) were compared at three cemeteries (Table 2). The age of the tombstones ranged over 108 yr and included a mix of veteran and non-veteran stones. The overall averaged difference between right and left sides at individual cemeteries is consistently an order of magnitude below the experimental error of ~1 mm.

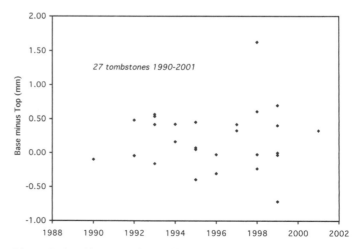

Figure 4. Graphic comparison of bottom-minus-top measurements of 27 relatively new tombstones (1990–2001). All but one of the tombstones had <±1 mm difference in thickness.

The standard deviation at each cemetery was also under 1 mm. Around three quarters of the tombstones had side-to-side differences of <1 mm, and ~3% had differences >2 mm.

If original thickness differences of the older tombstones (including non-veteran stones) were also on the order of 1 mm, then very low weathering rates (on the order of 1 mm/100 yr or less) of individual tombstones cannot be measured precisely using this method, unless tombstones are significantly older than 100 yr. Where measured recession rates are >1 mm/100 yr, they can be reported more precisely. The possibility that the base and top thicknesses were originally different complicates measurement of tombstones of all ages.

2. The Tombstone Base Remains Sufficiently Unweathered for Decades to Centuries

The base-minus-top measurement requires that tombstone bases can provide an accurate estimate of the original thickness (Meierding, 1981). The maximum age for which this method is viable at any cemetery will be constrained by the oldest age of tombstones with bases that are sufficiently unweathered. Any attempts to correct back to the original thickness using weathered bases are unlikely to yield results accurate to within 1 or 2 mm.

Visible weathering effects, especially loss of polish, rounding of cut corners, and color alterations may be enough to identify stones with bases that are too weathered to use. Variables that might influence the maximum age of sufficiently unweathered bases include intensity of precipitation events, biological impacts, pH of rainfall, height of tombstone, design and placement of the base, intensity of direct interaction with gases in the atmosphere, and others. Stones with visibly weathered bases were avoided in this study, so there is no comparison data to directly test the visual selection method.

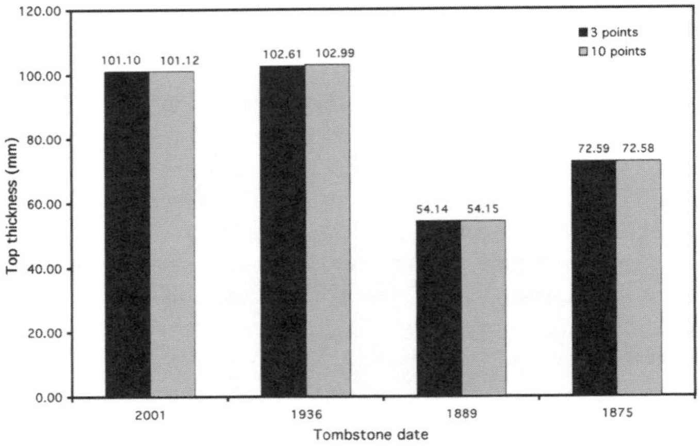

3. Weathering is Even across the Top of the Tombstone

This assumption is not verified by the measurements reported here, but it does appear to be possible to achieve a consistent average thickness measurement using three measurements across the top. Visual inspection reveals many weathering effects that are not evenly distributed, including cracking, rounding of outside corners, spalling, etc. Based on measurements of four tombstones with a 126-yr age range, three measurements of the top, as described in the methods section above, appear to provide an average thickness that is almost the same as 10 measurements (Fig. 5). The validity of using three as opposed to 10 measurements does not appear to alter with the age of the tombstone. Of course, there will be exceptions, for example where deep spalling or cracking has affected a small area of the tombstone top.

If there is a significant aspect effect in weathering, it is not apparent in the data presented here. Most of the tombstones measured in southwestern Montana face east, so that the "left" side is to the south and the "right" side is to the north, and it is easy to look for an aspect bias by comparing the right-left measurements

TABLE 2. RIGHT-LEFT DIFFERENCES IN TOMBSTONE BASE THICKNESSES

	Helena	Bozeman	Butte Sisters
Number of tombstones	68	49	57
Average right minus left (mm)	−0.09	0.16	−0.23
Standard deviation	0.84	0.94	0.92
Range	1.05 −2.01	3.07 −1.95	3.01 −2.02
% (n) with difference >1mm	26% (18)	24% (12)	19% (11)
% (n) with difference >2 mm	1.5% (1)	4.1% (2)	3.5% (1)

Note: Right = north-facing; left = south-facing aspect.

Figure 5. Test to compare the average thickness of tombstone tops, as determined by three or ten measurements. The three points were top left, top right, and top center. The ten points were roughly evenly spaced across the top. Tombstones were selected to be of different ages and of visually different degrees of weathering

of top (weathered) thickness. In fact, a right-left comparison of top measurements at all sites (Table 3) shows even less right-left difference than base measurements. The overall average difference between right and left sides at individual cemeteries is consistently two orders of magnitude below the experimental error of ~1 mm. The standard deviation at each cemetery was well under 1 mm. More than three quarters of the tombstones had side-to-side differences of <1 mm, and fewer than 2% had differences >2 mm. This result also supports the finding that three top measurements appear to be sufficient to characterize top thickness.

4. Surface Recession at the Top is Measurable and Acceptably Precise

Obviously, if the amount of weathering is not greater than the original (manufactured) thickness difference, it is not measurable by this method. In addition, the accuracy of the engineering

TABLE 3. RIGHT-LEFT DIFFERENCES IN TOMBSTONE TOP THICKNESSES

	Helena	Bozeman	Butte	Dillon
Number of tombstones	70	71	113	145
Average right minus left (mm)	−0.05	−0.05	0.05	0.00
Standard deviation	0.80	0.80	0.81	0.72
Range	1.83	1.83	2.59	3.92
	−1.99	−1.99	−2.01	−1.34
% (n) with difference >1mm	22.8% (16)	22.5% (16)	18.6% (21)	11.0% (16)
% (n) with difference >2 mm	0	0	2.7% (3)	2.1% (3)

Note: Right = north-facing; left = south-facing aspect.

calipers (0.01–0.02 mm) provides a constraint for measurement of surface recession by any method. For example, Cooke et al. (1995) noted that even a relatively large increase in weathering rate of 0.3 microns/year could take 30 yr to accumulate a measurable difference.

Another important concern is the precision of measurements. Precision is essential if very small differences are significant and especially if the rates determined by this method are to be extrapolated back to time periods orders of magnitude longer that the measured historical framework. In this study, precision was tested using a group of students in an upper-division geomorphology class. The comparison between different workers is important for a method that should be broadly applicable. Three different students took two separate measurements of four different tombstones. The precision of this method, when applied by different workers, appears to be a few tenths of one mm or better (Fig. 6). Periodic checks of operator variance in the field are usually within that range, although sometimes differences are up to 1 mm. Livingston and Baer (1990, p. 85) summarized precision of reported measurement (with calipers) of various parameters of tombstone recession to be "on the order of 0.1 mm."

The other question, then, is whether any particular cemetery environment includes tombstones that are old enough to have >1 mm surface recession. Feddema and Meierding (1987) suggested that recession cannot be measured until crystals are removed from the stone. Their data from Pennsylvania suggested that this process takes ~25 yr in urban areas to 60+ years in rural areas.

Work in Montana includes marble tombstones covering ~110+ years at four sites (Fig. 7). Based on >1000 measurements, this study predicts that the mean surface recession rates will be <2 mm/100 yr at rural sites in dry climates (Dillon), which is near the minimum measurable by this method. In urban non-industrial and urban industrial locations, the recession rates are >2 mm/100 yr. This range of 100-yr recession rates is comparable to those measured in other locations in the United States (Table 4).

At all four cemeteries, the largest measured maximum recession was not on the oldest stones (Fig. 7). The effect is most obvious at Butte, where the measured peak recession occurs at ca. 1900 and drops off remarkably after that. One way to interpret this result is to suggest that the older stones had already been thinned by weathering at the base that was not noted during field measurements.

5. Weathering over the Measured Period of Time is Dominated by Loss of Thickness of the Marble Slab

Although there are many observed effects of weathering on marble tombstones (crystal dissolution, whole-crystal loss, spalling, cracking, surface roughening, corner rounding, loss of inscription legibility, relief of lead lettering, etc.), most of them are related to loss of stone thickness through time. Some exceptions include discoloration and bending, which were not measured. Although the spread of data on individual tombstones is large in southwest Montana, this study records tombstone tops generally thinning through time with respect to bases at all cemeteries.

The data reported here appear to also indicate measurable expansion of some exposed top surfaces with respect to bases, especially in early stages of weathering. In Figure 7, both the Butte (7A) and Dillon (7D) graphs show increasing numbers of tombstones with tops thicker than bases with age, reaching a maximum in stones around 50 yr old. An increasing concentration of tombstones with negative base-minus-top measurements with increasing age would not be expected as a result of random placement of stones with original thickness differences.

Tombstone expansion may indicate that the stones are experiencing microscale loosening of crystals by thermal expansion-contraction cycles, frost wedging, growth of new minerals, release of stress related to quarrying, or other stresses, which are more intense at the exposed top surface of the stone. Expansion effects might continue to skew measured average recession rates until they are overwhelmed by recession effects—for example, when loosened crystals dissolve or begin falling out of the structure in large numbers. This process was mentioned briefly by Livingston and Baer (1990, p. 87) and is being elucidated by laboratory research on the early decay responses of stone subjected to repeated cycles of stress (e.g., Warke, 2002).

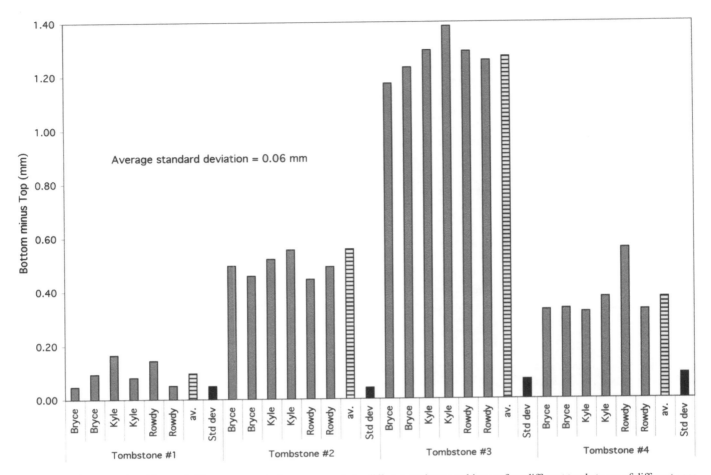

Figure 6. Precision test data. The graph illustrates repeated measurements by different students working on four different tombstones of different ages.

Data analysis, such as a scatter graph, indicates at what age thinning dominates in a particular cemetery, which helps constrain the minimum age for which the method is useful for determining average recession rate.

6. Surface Recession is Primarily a Linear Process because Thickness Loss is a Direct and Relatively Consistent Function of Exposure Time

Most researchers who have measured surface recession by whatever method describe a linear rate of weathering with time and report their data as average surface recession rates per year or per 100 yr (e.g., Matthias, 1967; Rahn, 1971; Feddema and Meierding, 1987; Meierding, 1993; Inkpen and Jackson, 2000). However, many authors, including the same ones who report their overall data as linear functions with time, also provide observations of the complexity of the processes involved, which supports a nonlinear interpretation of weathering rates. Schreiber and Meierding (1999, p. 180) note that although they assumed uniform weathering rates in their calculations, "it is virtually certain that rates have changed over time as pollution concentrations

have changed." Winkler (1987) and several others describe the apparently chaotic effects of cracking, crack-corrosion, spalling, etc. Feddema and Meierding (1987, p. 148) report that within-cemetery variance for 30 randomly selected stones increased from little-weathered (standard deviation 0.5 mm 100 yr^{-1}) to severely weathered (standard deviation 1.6 mm 100 yr^{-1}), "mostly due to different weathering rates between stones over time." They also note that if pollutant concentrations vary over time, then the weathering rate will not be linear and that initial weathering will appear slow because the amount of recession is not measurable at first (see above). A few time-series analyses have shown that weathering rates may be measurably different at different times in the same cemeteries (e.g., Dragovich, 1981; Husar et al., 1985).

Graphed primary data on this subject have been relatively scarce in the literature until recently; most published data are reported as mean surface recession (in mm or micrometers) per year or per hundred years or some other secondary form. Although this kind of information is very useful, it may also construct a façade around the true variability of the processes involved. The data presented here in "raw" form (Fig. 7) clearly demonstrate

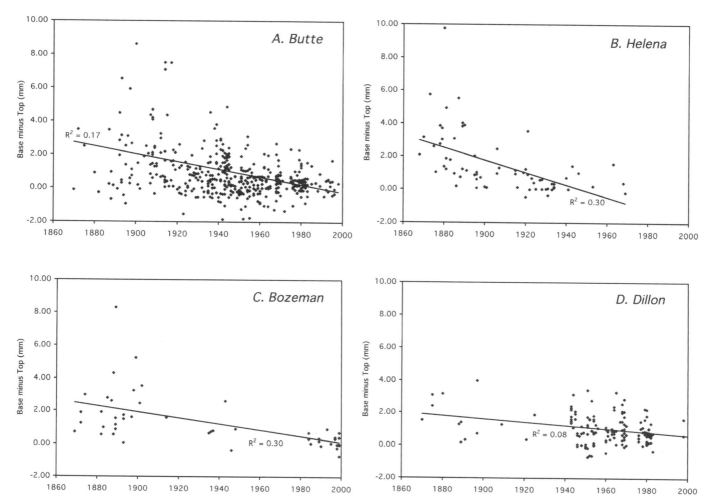

Figure 7. Surface recession data for individual tombstones plotted against date of tombstone for cemeteries at (A) Butte, (B) Helena, (C) Bozeman, and (D) Dillon. Visual inspection of the graphs shows large variability between tombstones of the same age in each cemetery. Low r² values for the linear regression line at each site indicate low predictive value for the linear model.

TABLE 4. COMPARISON OF WEATHERING RATES OF VERTICAL
MARBLE TOMBSTONES IN THE UNITED STATES

Weathering rate (mm/100 yr)	Location	Cultural setting Urban/Industrial	Rural	Reference
2.0	SW Montana	x		This study (urban average)
<1.0	SW Montana		x	This study
5?	Philadelphia	x		Baer and Berman (1983)
3.5?	New York City	x		Baer and Berman (1983)
2.8	Connecticut		x	Rahn (1971)
3.5	Philadelphia (center)	x		Feddema and Meierding (1987)
<0.5	Philadelphia (suburbs)		x	Feddema and Meierding (1987)
1.4	Upper Midwest USA		x	Meierding (1993)
0.9	Louisville, Kentucky	x		Meierding (1993)
3.0	SW Pennsylvania	x		Schreiber and Meierding (1999)
1.1	SW Pennsylvania		x	Schreiber and Meierding (1999)
>3	Chicago	x		Winkler (1987)

that surface recession of marble tombstones at the four reported sites is not a strictly linear process with time, at least as viewed from the combined record of lithologically similar stones at the same cemeteries. Several important observations include:

- Linear regression lines do not all cross the *x*-axis at time zero (present). A nonzero intercept implies some aspect of nonlinearity of weathering rates with time. Inkpen (1998) and Inkpen and Jackson (2000), measuring recession of marble with respect to lead lettering, noted the same effect in their published primary data. They argued that the problem might be because initial slow or rapid weathering is followed by weathering that approaches linearity with time. Inkpen and Jackson (2000) affirm that linear regression is an appropriate model for their 40-to 12-year-old tombstones. Their r² values were much higher than those of this study (averaging 0.64 for six sites). Pope et al. (2002) reviewed recession rate studies and found no agreement in the geomorphic literature on the subject of weathering rate change with time, with arguments made for both accelerating and decelerating rates. In addition, in the southwest Montana data, the youngest tombstones do not cluster tightly around the zero line, probably because they were not exactly the same original thickness at the top and base.
- The degree of weathering generally increases with age, but the data is scattered over a very broad area, resulting in low r² values. The causes of this broad scatter may be nonlinear processes. In addition, initially small differences are magnified by weathering processes (linear or nonlinear) as the stones age. Stones with initially thicker tops than bases will also provide scatter as they weather.
- Maximum surface regression at an individual site may approximate a linear trend. The dashed line on Figure 8 follows the most weathered stone for each decade at Butte (1900–1980). This "maximum" weathering line still probably underestimates the maximum surface recession; the oldest and most weathered tombstones were not measured because the bases had obviously begun to weather.
- At some sites, the number of tombstones that are thicker at the top appears to increase during the first 40–60 yr of exposure (Fig. 8). The data may document an initial expansion stage.

The literature is full of documented observations of processes that might be expected to produce nonlinear weathering results over time, but the difficulty of using nonlinear models to describe and predict change, and the usefulness of mean recession values, has probably kept the assumption of overall linearity alive. Over certain age ranges and at some sites, the cumulative effect of all weathering processes may actually be linear (e.g., Inkpen and Jackson, 2000). Several authors have recently suggested that surface recession rates of stone could be better described as episodic or even chaotic (Viles, 2002; Pope et al., 2002; Warke, 2002). In a quantitative study of acid rain damage to carbonate stone, Reddy (1988) offered a graph of rainfall amount versus $CaCO_3$ recession that is distinctly nonlinear.

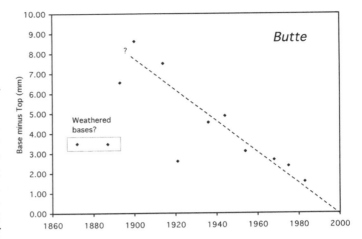

Figure 8. "Eyeball" approximation of a linear maximum recession rate at Butte, using the tombstone with the largest base-minus-top difference per decade from 1900 to 1980 (excluding the 1920s). The origin was artificially set. After 1900, the maximum recession falls off rapidly, suggesting that the bases of these older tombstones were already too weathered to provide original thickness estimates.

The work reported here compares stones of similar lithology and shape, oriented in about the same direction in the same and different cemeteries, but did not track individual stones through time. Lab and field experiments designed to follow the deterioration of individual stones through an extended time may eventually provide that information (e.g., Yerrapragada et al., 1996; Butlin et al., 1992; the U.S. Bureau of Standards Stone Wall Test that has exposed a wall of various stones since 1948 [Stutzman and Clifton, 1997]). Similarly, a tighter control on tombstone lithology might produce more linear results. Nevertheless, the nonlinear or chaotic possibilities of this process deserve more analysis.

CONCLUSIONS

Measurement of the surface recession of marble tombstones by weathering is a well-documented procedure. Like any other scientific procedure, its usefulness is constrained by underlying assumptions, which deserve constant reevaluation.

This research emphasizes that, although surface recession of marble tombstones (and presumably other similar cultural stones) increases with time, the function is complex and probably nonlinear. An early result of weathering may be expansion of the top of the stone relative to the base. This effect appears to begin before loss of polish and continue for several decades.

The data reported here support previous indications that the weathering loss for an individual stone of any particular age or environment is not predictable by this method. However, for each environment there may be a maximum expected weathering loss that could be predicted to assist the design and preservation of stone monuments. Surface recession data for any particular cemetery can also be used to define the minimum and maximum tombstone ages for which the method described here will provide the most useful information on surface material loss.

ACKNOWLEDGMENTS

Working in cemeteries can be depressing. I greatly appreciate the patient, good-natured contribution of the students in my surficial processes class, 1999, who helped get this research underway, and Tom Satterly and Pansy Bradshaw, who provided field assistance for continued measurements since 2000.

REFERENCES CITED

Baer, N.S., and Berman, S.M., 1983, Marble tombstones in national cemeteries as indicators of stone damage: General methods: Preprints, 76th Annual Meeting, Air Pollution Control Association, p. 19–24.

Butlin, R.N., Coote, A.T., Devenish, M., Hughes, I.S.C., Hutchens, C.M., Irwin, J.G., Lloyd, G.O., Massey, S.W., Webb, A.H., and Yates, T.J.S., 1992, Preliminary results from the analysis of stone tablets from the national materials exposure programme (NMEP): Atmospheric Environment, v. 26B, p. 189–198.

Cooke, R.U., Inkpen, R.J., and Wiggs, G.F.S., 1995, Using gravestones to assess changing rates of weathering in the United Kingdom: Earth Surface Processes and Landforms, v. 20, p. 531–546.

Dragovich, D., 1981, Weathering rates on marble tombstones at a Sydney cemetery: Institute for the Rate of decay of marble in laboratory and outdoor exposure: Journal of Materials in Civil Engineering, v. 1, no. 2, p. 73–85.

Dragovich, D., 1987, Measuring stone weathering in cities: surface reduction on marble monuments: Environmental Geology and Water Sciences, v. 9, p. 139–142.

Dragovich, D., 1991, Marble weathering in an industrial environment, eastern Australia: Environmental Geology and Water Sciences, v. 17, p. 127–132.

Feddema, J., and Meierding, T., 1987, Marble weathering and air pollution in Philadelphia: Atmospheric Environment, v. 21, no. 1, p. 143–157, doi: 10.1016/0004-6981(87)90279-4.

Gauri, K.L., Popli, R., and Sarma, A.C., 1983, Effect of relative humidity and grain size on the reaction rates of marble at high concentrations of SO_2: Durability of Building Materials, v. 1, p. 209–216.

Gauri, K.L., Kulshreshtha, N.P., Punuru, A.R., and Chowdhury, A.N., 1989, Rate of decay of marble in laboratory and outdoor exposure: Journal of Materials in Civil Engineering, v. 1, no. 2, p. 73–85.

Geikie, A., 1880, Rock-weathering as illustrated in Edinburgh church yards: Proceedings of the Royal Society of Edinburgh, v. 10, p. 518–532.

Husar, R.B., Baer, N.S., and Patterson, D.E., 1985, Deterioration of marble: A retrospective analysis of tombstone measurements in the New York City area: Springfield, Virginia, National Technical Information Service, PB85-174134.

Inkpen, R.J., 1998, Gravestones: Problems and potentials as indicators of historic changes in weathering, *in* M.S. Jones and R.D. Wakefield, eds., Aspects of stone weathering, decay and conservation: Stone Weathering and Atmospheric Pollution Network '97: London, Imperial College Press, p. 16–27.

Inkpen, R.J., and Jackson, J., 2000, Contrasting weathering rates in coastal, urban and rural areas in southern Britain: Preliminary investigations using gravestones: Earth Surface Processes and Landforms, v. 25, p. 229–238, doi: 10.1002/(SICI)1096-9837(200003)25:3<229::AID-ESP52>3.0.CO;2-Y.

Livingston, R.A., and Baer, N.S., 1990, Use of tombstones in investigation of deterioration of stone monuments: Environmental Geology and Water Science, v. 16, no. 1, p. 83–90.

Matthias, G., 1967, Weathering rates of Portland Arkose tombstones: Journal of Geological Education, v. 15, p. 140–144.

Meierding, T.C., 1981, Marble tombstone weathering rates: A transect of the United States: Physical Geography, v. 2, no. 1, p. 1–18.

Meierding, T.C., 1993, Inscription legibility method for estimating rock weathering rates: Geomorphology, v. 6, p. 273–286, doi: 10.1016/0169-555X(93)90051-3.

Pope, G.A., Meierding, T.C., and Paradise, T.R., 2002, Geomorphology's role in the study of weathering of cultural stone: Geomorphology, v. 47, p. 211–225, doi: 10.1016/S0169-555X(02)00098-3.

Rahn, P.H., 1971, The weathering of tombstones and its relationship to the topography of New England: Journal of Geological Education, v. 19, p. 112–118.

Reddy, M.M., 1988, Acid rain damage to carbonate stone: A quantitative assessment based on the aqueous geochemistry of rainfall runoff from stone: Earth Surface Processes and Landforms, v. 13, p. 335–354.

Schreiber, K.V., and Meierding, T.C., 1999, Spatial patterns and causes of marble tombstone weathering in western Pennsylvania: Physical Geography, v. 20, p. 173–188.

Skoulikidis, Th., and Charalambous, D., 1981, Mechanism of sulphation by atmospheric SO_2 of the limestones and marbles of the ancient monuments and statues: British Corrosion Journal, v. 16, p. 70–77.

Stutzman, P.E., and Clifton, J.R., 1997, Stone exposure test wall at NIST, *in* Labuz, J.F., ed., Degradation of natural building stone: American Society of Civil Engineers Geotechnical Special Publication 72, p. 20–32.

VanGemert, D., Ulrix, E., and Viaene, W., 1988, Physico-chemical erosion of limestone: A micro-macro approach, *in* Marinos, P.G., and Koukis, G.C., eds., Engineering geology of ancient works, monuments and historical sites: Rotterdam, Balkema, p. 805–808.

Viles, H.A., 2002, Is stone decay chaotic? Reflections from studies of the limestone cultural heritage of Oxford, UK: Geological Society of America Abstracts with Programs, v. 34, no. 6, p. 89.

Warke, P.A., 2002, Complex weathering effects on the durability of building stone: Geological Society of America Abstracts with Programs, v. 34, no. 6, p. 89.

Wakefield, R.D., and Jones, M.S., 1998, An introduction to stone colonizing micro-organisms and biodeterioration of building stone: Quarterly Journal of Engineering Geology, v. 31, p. 301–313.

Winkler, E.M., 1987, Weathering and weathering rates of natural stone: Environmental Geology and Water Science, v. 9, no. 2, p. 85–92.

Yerrapragada, S.S., Chirra, S.R., Jaynes, J.H., Li, S., Bandyopadhyay, J.P., and Gauri, K.L., 1996, Weathering rates of marble in laboratory and outdoor conditions: Journal of Environmental Engineering, v. 122, p. 856–863, doi: 10.1061/(ASCE)0733-9372(1996)122:9(856).

MANUSCRIPT ACCEPTED BY THE SOCIETY 19 JANUARY 2005

Geological Society of America
Special Paper 390
2005

Petra revisited: An examination of sandstone weathering research in Petra, Jordan

Thomas R. Paradise*

Department of Geosciences and the Fahd Center for Middle East and Islamic Studies,
University of Arkansas, Fayetteville, Arkansas 72701, USA

ABSTRACT

Petra, Jordan, was an important crossroads city occupied during the Nabataean and Roman eras. This paper presents a review of a series of studies conducted from 1990 to 2003 that scrutinized both lithologic (intrinsic) and climatic and anthropogenic (extrinsic) weathering influences on the Paleozoic sandstones in Petra, an ideal environmental "laboratory" for the study of weathering features, causes, and rates. However, these structures, which have been stable since their creation two thousand years ago, are deteriorating at an accelerated rate due to natural and human-induced stone decay processes.

Comprehensive measurements of surface recession were made in Al-Khazneh, Petra's most celebrated tomb; the Roman Theater; and the Anjar Quarry above this abandoned city. Surface recession rates for sandstone in the Roman Theater were determined to range from 15 to 70 mm/k.y. on horizontal surfaces to 10–20 mm/k.y. on vertical surfaces. Higher iron and silica contents of sandstone matrix were found to decrease overall sandstone weatherability, while calcareous matrix components were found to increase deterioration in areas that receive >5500 MJ/m^2/yr of solar radiation. Moreover, when iron matrix concentrations exceed 4%–5% (by weight), original stonemason dressing marks are still clearly evident, indicating a nearly unweathered state in 2000 yr. Visitors to Petra have dramatically increased from 100,000 (1990) to 350,000 (1998). Large (and typical) tourist groups entering the chamber of Al-Khazneh were found to raise interior relative humidity levels from 20% to 50%, and interior surfaces have dramatically receded due to visitor touching, as much as 40 mm in less than 50–100 yr.

Keywords: sandstone deterioration, Petra, Nabataean architecture, weathering, sandstone lithology, relative humidity, anthropogenic weathering, aspect, surface recession, microclimate, stone decay, rock properties, matrix constituents, conservation strategies.

INTRODUCTION

The ruined crossroads city of Petra in Jordan's southern desert lies deep in a valley surrounded by steep, impassable sandstone walls and winding, faulted gorges (Fig. 1). However, it is the dis-

tinctive architecture rather than its spectacular setting that makes this ancient city so important. Although archaeological evidence indicates occupation in the Valley of Petra since 7000 BCE, it was the indigenous Arab residents, the Nabataeans, and their Roman visitors who brought great power to the region (Taylor, 2002). They worked the valley walls into simple caves and elaborately carved tombs and structures, hewn directly from the reddish

*E-mail: paradise@uark.edu

Paradise, T.R., 2005, Petra revisited: An examination of sandstone weathering research in Petra, Jordan, *in* Turkington, A.V., ed., Stone decay in the architectural environment: Geological Society of America Special Paper 390, p. 39–49, doi: 10.1130/2005.2390(05). For permission to copy, contact editing@geosociety.org. ©2005 Geological Society of America.

brown and yellowish sandstone cliffs, many exceeding 50 m in height. Since then, climatic, lithologic, and human factors have been influential in the deterioration of Petra's architecture.

Weathering research has typically separated weathering factors into two categories: those affected by the characteristics of the material itself or intrinsic effects (i.e., lithologic constituents, fractures), and those affected by external influences or extrinsic effects (i.e., climate, human contact). The weathering of Petra's sandstone architecture has been analyzed in terms of recessional surface features related to variability in rock composition and/or caused by weathering influences such as insolation, running water, or human touch. This research indicated that extrinsic influences are as significant as intrinsic factors in determining the rate of sandstone deterioration. This paper presents a review of a series of studies conducted in Petra between 1990 and 2003 that attempted to answer the following questions (Paradise, 1995, 1999a, 1999b, 2000, 2002):

1. What are the intrinsic factors that have controlled the rate and nature of sandstone weathering in Petra?

2. What are the extrinsic factors that have controlled the rate and nature of sandstone weathering in Petra?

3. What has been the rate of sandstone surface recession in Petra, and how has this changed?

Weathering studies of sandstone architecture in arid environments are relatively rare. Early observations on stone and architectural deterioration in the Near East were made by Herodotus (ca. 450 BCE), Strabo (ca. CE 10), Pliny (ca. CE 50), J.L. Stephens (ca. 1830) and R.F. Burton (1879). In the early twentieth century, Bryan (1922, 1928) and Blackwelder (1929), for example, discussed many of the processes responsible for the development of unique sandstone weathering features. These were some of the first Western works that addressed sandstone weathering conceptually rather than simply describing weathering features (e.g., tafoni).

Subsequent research in arid regions established important relationships between weathering and various influences including lichen overgrowth (i.e., Jackson and Keller, 1970; Jones et al., 1980; Paradise, 1997), tafoni development (i.e., Mustoe, 1983), case hardening and permeability (i.e., Conca and Rossman, 1982; Pfluger, 1995), salt content (i.e., Smith and McGreevy, 1988; Young, 1987), insolation and moisture availability (i.e., Fahey and LeFebure, 1988; Robinson and Williams, 1992; Paradise, 1995, 1998). This research indicates that sandstone principally weathers in two ways. Since sandstone is comprised of clasts (particles) within a matrix (binder), either the clast fractures or dissolves and then falls out or the matrix fractures or dissolves to release the clast. Both weathering types represent the processes of disaggregation that produce loose sand as the by-product of sandstone deterioration, the source of many of the sand dunes throughout the Near East and North Africa.

METHODOLOGY

Historical structures and monuments represent a valuable resource for the study of weathering influences and recession rates. Limestone and marble tombstones, for instance, have been used for weathering analysis since Geikie's work (1880) in Scottish cemeteries. Marble and limestone weathering using structures and objects have been studied by Rahn (1971), Hoke (1978), Meierding (1981), Klein (1984), Dragovich (1986), Neil (1989), and Trudgill (1989). Granite architectural decay has been studied by Winkler (1965), Amoroso and Fassina (1983), and Smith and Magee (1990), while other materials used in construction, such as arkose (Matthias 1967) and calcarenite (Spencer 1981), have also been examined in this context. The research reviewed here used similar techniques whereby the known relative or absolute date of the structure (or objects) and the known original surface allows determination of the change due to weathering-induced recession since its construction. This difference between the original and current rock surfaces then facilitates investigation of the intrinsic and extrinsic weathering factors, and their relationships, affecting rock surfaces. In Petra, the difference between the current weathered sandstone surface and the original dressed Nabataean or Roman surface represents how much of the sandstone has receded due to weathering and erosion over 2000 yr.

Much stone decay research has been conducted in Europe, Australia, and the United States, where friable sandstones are not as commonly found in construction as the more resistant and accessible stone building materials such as granite, limestone, or marble. This is why sandstone is an often under-studied building material, potentially valuable for weathering studies in arid regions where its use is common because of its abundance and where its weatherability is decreased due to reduced precipitation—an important factor in sandstone deterioration (Paradise, 1995; Pope et al., 2002). This paper presents three case studies that utilized surface recession measurement strategies to understand the relationships between sandstone weathering and lithology, aspect, and anthropogenic influences in Petra, Jordan.

WEATHERING AND LITHOLOGY

An extensive study was conducted on the Roman-style Theater of Petra (Paradise 1995, 1999a), a huge arena that may have seated up to 10,000 people (Fig. 2). Although it was carved before full Roman occupation in 106 CE, it was hewn out of the sandstone cliffs according to the canon of the great Roman engineer, Marcus Pollio Vitruvius, using the highest Roman engineering and construction standards of the period. These early requirements for theater and building construction were so standardized that the level of the original surfaces can be estimated from the current receded surfaces. Five hundred locations were examined across the theater for intrinsic factors (variations in rock composition and particle size) and extrinsic influences (sunlight angle, lichen coverage, slope, and surface temperature).

The theater research yielded important relationships. First, it was found that sandstone recession is accelerated by insolation (where temperatures exceed 50 °C), by carbonate content, and by wide spacing of sandstone sand grains (high matrix-to-clast ratio). High insolation receipt on sandstone surfaces was found on

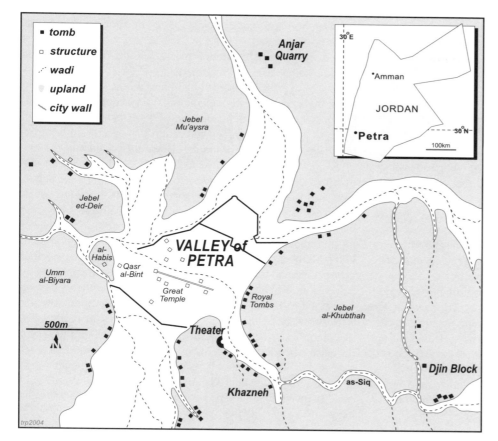

Figure 1. Map of the Valley of Petra, Jordan.

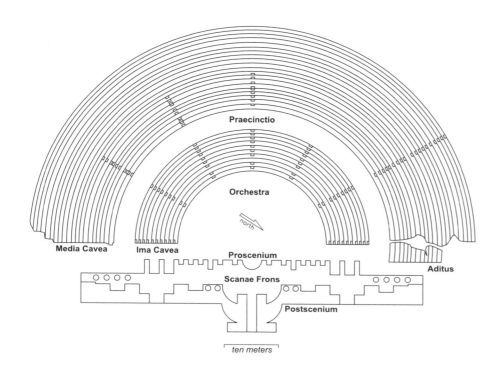

Figure 2. Schematic diagram of Petra's Roman Theater, which was hewn from the surrounding sandstone cliff s during the first century. Note: only the *ima* and *media cavea* were used in the study since the *summa cavea* appears to have not been hewn according to the more precise Vitruvian canon (making surface recession estimation more difficult).

southerly aspects, which received >5500 MJ/m²/yr. Rock properties, it was discovered, could be interpreted on the basis of sandstone color, since paler calcium-rich and/or high-matrix sandstones are often softer and more friable than darker colored, iron-rich sandstones. Second, other relationships were found in which lithologic factors decreased weathering. It was discovered that iron oxides (Fe_2O_3, FeO_2) and silica (SiO_2) in the sandstone matrix dramatically decreased weathering. Simple observations supported these findings where the original Roman stonemason tool marks were obvious atop iron-rich sandstone (dark reddish) while large weathered cavities had developed on the rock areas that were iron-poor (pinkish-beige to white). In fact, it was found that when matrix iron concentrations exceeded 4% (by weight), dressing marks appeared so fresh and deep that little or no sandstone deterioration was apparent. Inferential statistical analysis found that the iron and silica proportions of the sandstone matrix (with combined r^2) explained 50% of all deterioration in the Roman Theater (Table 1).

The Theater study not only identified important weathering influences, it also established minimal weathering rates for the Petra sandstone. Based on surface measurements taken across the Theater, minimum rates of surface recession were estimated from 15 to 70 mm/k.y. on horizontal surfaces to 10–20 mm/k.y. on vertical surfaces. Lower vertical slope recession may be attributed to decreased occurrence of standing water and saturation (and diminished wetting and drying cycles), but this has also been produced by recent and ancient abrasion from spectators and tourists.

In addition, this study established a previously unknown hierarchy of weathering processes responsible for sandstone decay. Principal component statistical analysis (PCA) was used to explain the relative importance of the various agents weathering sandstone in an arid environment like Petra. It was found that general rock composition (measured through backscatter scanning electron microscope [SEM]) was the most important single influence (25%), followed by the effects of iron concentration (17%) and climatic influences like sunlight and moisture (12%), in affecting the weathering of stone architecture. These findings emphasize the importance of lithology in understanding and predicting stone decay rates, specifically in Petra and generally in arid regions. Moreover, such a hierarchy is vital in grasping the comparative controls on sandstone deterioration and the possible priorities needed in conservation applications and research.

Deterioration of the sandstone of the Roman Theater has recently accelerated. When the Theater was first examined for this research in 1990, at least 15%–20% of the Theater displayed original stonemason dressing marks made ca. 2000 yr B.P. However, it was recently observed that these marks are disappearing at a faster rate, especially on horizontal surfaces affected by visitor contact (foot-tread). In 2001, only 5%–10% of the theater still exhibits stone dressing—an indication that sandstone weathering of the theater is accelerating. Since this change in the rate of decay cannot be attributed to intrinsic changes, or a change in climate, it must be attributed to changes in tourism, such as increased foot-tread by new styles of footwear. Also, this accelerated surface recession is especially evident in the areas near the orchestra, *praecincterae*, and *itinerae*, parts of the theater most commonly visited by tourists and tour group operators in Petra. Many more people are visiting Petra, promoted by cavalier government and tour policies, and erosion has been enhanced by newer, soft, gripping shoe soles that increase the relative friction between the visitors' feet and the sandstone. Therefore, however sensitive visitors may be in their interaction with monuments and landscape, visitor-accelerated weathering in Petra will only decrease when tourism decreases, shoe soles become less abrasive, and/or visitor access is restricted. One recommendation is to limit access of theater visitors to the orchestra and *praecincterae* (concentric walkways), with minimal access given to the main theater since the commonly observed practice of jumping between theater seatbacks drastically increases surface deterioration and recession. This practice of jumping between seatbacks as a quick way to climb to the theater's *summa cavea* for a great view of Petra is typically recommended by most tour leaders.

WEATHERING AND ASPECT

Research since Blackwelder (1933) has indicated that sunlight, if not a direct cause of weathering, may have a significant influence on stone weathering. Roth (1965) confirmed Griggs' (1936) earlier work on the influences of temperature but emphasized the importance of sunlight as an overall weathering agent. Later, Smith (1977) established that rock (limestone) temperatures vary greatly with time of year and aspect, a direct influence of insolation. Research followed that explained a

TABLE 1. SANDSTONE SURFACE RECESSION RELATIONSHIPS
TO LITHOLOGIC CONSTITUENTS (r^2)

Density (gross g/cm²)	−0.714 ($n = 14$)	−0.416 ($n = 202$)	−0.403 ($n = 202$)
Iron content (matrix %)	−0.714 ($n = 14$)	−0.282 ($n = 202$)	−0.289 ($n = 264$)
Silica content (matrix %)	−0.714 ($n = 14$)	−0.743 ($n = 14$)	−0.648 ($n = 202$)
Matrix (gross %)	0.507 ($n = 202$)	0.436 ($n = 202$)	0.377 ($n = 264$)
Calcium content (matrix %)	0.433 ($n = 202$)	0.373 ($n = 202$)	0.254 ($n = 264$)

Note: All correlations of determination are statistically significant at <0.01. These correlations of determination (r^2) represent the relations between the various lithologic variables and overall sandstone surface recession on surfaces of the *proscenium* (left), *ima cavea* (middle) and *media cavea* (right) of the Roman Theater.

number of interesting relationships between insolation and differential weathering related to albedo and conductivity (Kerr et al., 1984), salt and relative humidity (Sperling and Cooke, 1985), and increased heating-cooling cycles (Jenkins and Smith 1990; Paradise 2002).

More specifically, studies have examined topography, aspect, and moisture availability as surrogates for insolation-induced weathering. Sancho and Benito (1990) discussed weathering features in Spain; Robinson and Williams (1989, 1992) explained surface morphology (polygons, skins, gnammas) in France and Morocco; and Paradise and Yin (1993) addressed gnamma or solution pit development, size, and shape in the U.S. state of Georgia. These studies have explained the importance of aspect, heating-cooling cycles, and moisture availability in the acceleration of surface recession and/or weathering feature development, reinforcing the notion that received solar radiation is important in the acceleration of weathering. In fact, the research at Anjar,

the quarry from which Petra sandstone was sourced, found the largest recession features on southwest and southeast faces, supporting the importance of the tandem role of heating/cooling and wetting/drying cycles.

Among the sandstone cliffs of Petra are the ancient Nabataean quarries, including those high above Petra at Anjar, in addition to a number of unique carved "cubic blocky" sculptures called the Djin Blocks. The sandstone surfaces of the quarry and the blocks were studied as many of these hewn surfaces face different aspects, therefore making them ideal for observing the effects of sunlight on stone deterioration. The Nabataeans prepared these quarry surfaces by chiseling and dressing them into distinctive herringbone patterns (Shaer, 1997), a style not used by Romans stonemasons (Figs. 3 and 4). These dressed faces were chosen because their surfaces were carved at the same time (50 BCE to 100 CE), are vertical surfaces with easy access, and have not been modified or obscured since their exposure (Figs. 4 and 5). Using the original

Figure 3. An aerial view of Anjar Quarry, looking to the south toward the Valley of Petra

Figure 4. A Nabataean stonemason-dressed northern slope in Anjar Quarry.

Figure 5. North-facing sandstone exhibiting extensive lichen overgrowth near Siq al-Barid (darker color on lighter stone is *Lecanora sp.*), Petra.

Nabataean-dressed surfaces as baselines, recessional features like pits, channels, and cavities were located, mapped, and measured (Figs. 6 and 7).

Weathering features (i.e., tafoni, cavities) on the dressed quarry faces were measured, and a number of relationships were divulged. Northern-facing surfaces (340–020 °N) showed the least recession with 90% of the original stone-dressing visible and no recessional features exceeding 2 cm in any dimension. The discrete and measured features were then mapped where their dimensional values (width, depth) were related to the quarry wall aspects (Fig. 8). The relatively minor weathering observed on northern faces can be attributed to decreased weathering from lichen overgrowth (Fig. 9), since lichens are rarely found on other surfaces, and to the fact that less sunlight produces fewer wetting and drying cycles. Even though it has been widely accepted that lichen attachment can accelerate substrate weathering through rhizinal penetration and oxalic acid production, in Petra it was found that lichen overgrowth (*Lecanora sp.*) acts as a sandstone surface–consolidating agent, which decreases overall surface recession (Paradise 1995, 1999b). It was also observed that these northern faces rarely exhibited increased sheltering from rock overhangs or tree shading.

Southern faces (185 °N) displayed 40% of the original dressing with few recessional features larger than 15 cm in any dimension. The increased weathering on southern faces may be due to the greater amount of sunlight exacerbating daily heating and cooling cycles. Western to southwestern faces (230–250 °N) and eastern to southeastern faces (070–110 °N), however, displayed the greatest amount of deterioration and recession, with hardly any original Nabataean stone-dressing remaining (<10%), and with many cavities exceeding 20 cm across. This may be attributed to the optimal daily and yearly climatic regime experienced by these quarry faces, where temperatures are hot enough to expand the sandstone, causing disaggregation, and repeatedly moist conditions cycle the rock between hot and cool, thawed and frozen, wet and dry—cycles known to accelerate rock weathering (Fahey and Dagresse, 1984; Hall, 1986; Fahey and LeFebure, 1988). A similar distribution of tafoni related to aspect was observed on the Djin Blocks (Fig. 10).

These discoveries reinforce the conventional notion that the deterioration of sandstone is greatly accelerated from increased heating and cooling or wetting and drying cycles. However, it is now believed to be faster and more destructive than previously understood. In Petra, this research demonstrates the delicate balance of weathering between climate and lichen overgrowth on stone architecture.

WEATHERING AND VISITOR IMPACTS

Research conducted in the interior chamber of Petra's Al-Khazneh Tomb investigated human-induced changes in relative humidity and the effect on weathering rates. Prior research (and the research at the Anjar quarries) indicated that moisture availability and temperature fluctuation both influence stone deterioration. However, few studies on the direct effects of humans on

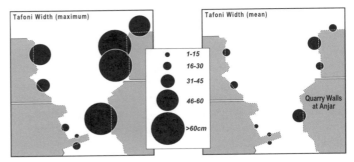

These thematic maps represent planimetric views of the Anjar quarry walls in Petra. The varying circle sizes represent the differing individual tafone dimension maxima and mean in centimeters. The influence of aspect is apparent as tafoni dimension and aspect are mapped.

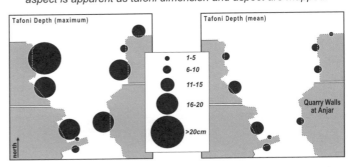

Figure 6. An illustration of the Anjar Quarry tafoni, showing surface recession variability related to aspect.

small-space humidity changes can be found. The Al-Khazneh temple or tomb was elaborately carved with classical elements (pediment, columns, entablature, etc.) and consists of a *propylaea* or large approaching staircase, two small exterior flanking chambers, and one primary chamber with three antechambers (interior volume ~2000 m³). This hewn structure was made famous in Spielberg's 1987 film "Indiana Jones and the Last Crusade" and is Petra's most popular visitor attraction, often holding more than one hundred tourists at a time (Figs. 11 and 12). Since 1998, environmental monitoring in the interior of Al-Khazneh has indicated that there is a strong relationship between large numbers of visitors in the tomb and a subsequent rise in relative humidity (unrelated to outside climatic fluctuations). The data indicate that the greatest increases in humidity occur when visitor groups exceeding 25–30 people remain within the tomb >5 min (Fig. 13). This is an important finding since many tour groups visiting Al-Khazneh in Petra consist of at least 30 people and remain >10 min in the inner chamber as tour group leaders explain the history (actual or contrived) of the site and Petra.

Prior studies have shown that increased moisture in restricted spaces can increase the production of surface salts (efflorescence), in-rock permeability, and moisture wicking and cause a general accelerated weathering and surface recession of sandstone due to particle disaggregation (Paradise, 1999b). Extensive research in arid regions (i.e., Egypt and Arizona) suggests that drier structures exhibit slower deterioration rates than

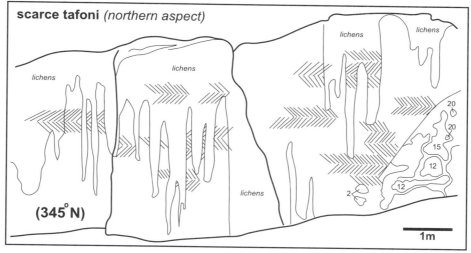

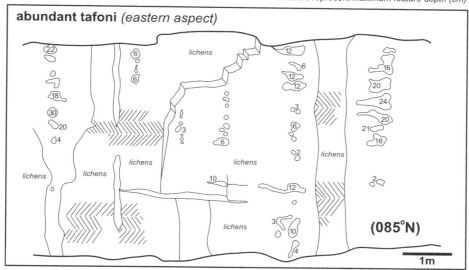

Figure 7. An illustration of the Anjar Quarry tafoni, showing that northern aspects (345 °N) display dramatically less surface recessional features such as tafoni, while eastern aspects (085 °N) exhibit more and markedly larger recessional features.

Comparative Sandstone Surfaces and Tafoni
Anjar Quarry, Petra

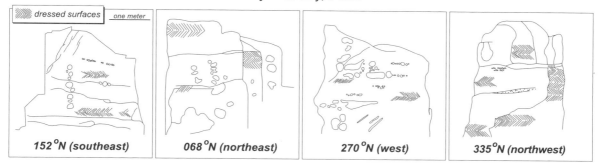

Figure 8. An illustration of tafoni development on four comparative sandstone surfaces in Anjar Quarry.

Tafoni Dimensions and Aspect
Anjar Quarry, Petra

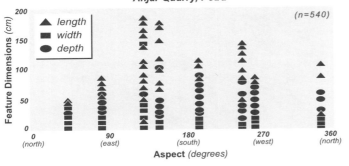

Figure 9. Tafoni dimensions related to aspect in Anjar Quarry.

Sandstone Surface Recession
Bab As-Siq Djin Block #5

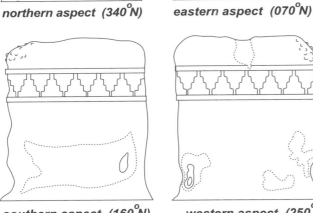

Figure 10. An illustration of tafoni development on a Djin Block showing differences in surface recession due to moisture and insolation variations on each side.

Figure 11. The Al-Khazneh Tomb façade, rising more than 35 m above a narrow defile called the Outer Siq. The Khazneh (or treasury) may have been an elaborate royal burial chamber, but is now the main attraction to Petra's thousands of yearly visitors.

Figure 12. The outer chamber of the Al-Khazneh Tomb. The sandstone's finely hewn detailing along the jambs, pilasters, mantle, and entablature, carved 2000 yr B.P., remains relatively unaltered, while most chamber interior detailing has deteriorated.

wetter ones (i.e., Emery, 1960). Precisely how this increased moisture regime contributes to accelerated deterioration in Petra, however, needs further study. So, as tourist numbers swell in Petra's chambers or tombs, interior humidity will rise, increasing moisture cycles, thus accelerating deterioration. Because Petra's visitors have grown from ~100,000 in 1990 to ~350,000 in 1998 (with single day attendance often nearing 4000 in 1999), continued monitoring of all environmental variables (external and internal) is recommended in order to evaluate carrying capacities and accessibility in this United Nations Educational, Scientific and Cultural Organization World Heritage Site (Fig. 14).

Finally, since the tomb chambers were hewn directly from the local sandstone cliffs, many of these interior surfaces still display the original stonemason dressing grooves from 2000 yr ago. In Al-Khazneh, for example, the four interior walls all exhibit original stone dressing marks. However, in the past decade, it has become increasingly obvious that many of these dressed grooves are rapidly deteriorated from 0.5 to 2.5 m above the tomb floor. This surface recession was mapped in 1999 using laser leveling devices to create a virtual netted surface model (using SURFER© visualization technology); it was found that across the 4 × 3 m tomb chamber wall surface, 0.53 m³ (526,000 cm³) of sandstone had eroded. This recent recession suggests that it is caused by visitor contact, such as touching and leaning. In fact, on the more than 100 occasions that Al-Khazneh was visited, local police, tourists, and tour guides were found leaning, touching, rubbing, and propping their feet against the chamber walls, producing small, loose sand piles at the base of the measured wall. Possible solutions to decrease interior humidity fluctuations may involve modification of interior microclimates (such as adding fans or dehumidifiers) or the restriction of in-tomb visitor numbers at any one time, with gaps between visitor groups long enough to permit the tomb chambers to re-stabilize to a naturally lower humidity. In addition, a solution to decrease wall abrasion by visitor contact may be to simply restrict or prohibit direct access to the tomb interior walls.

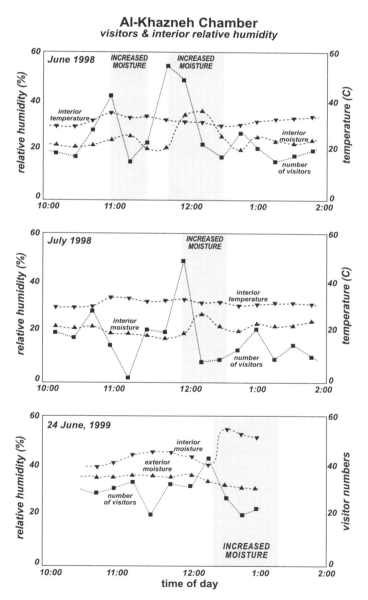

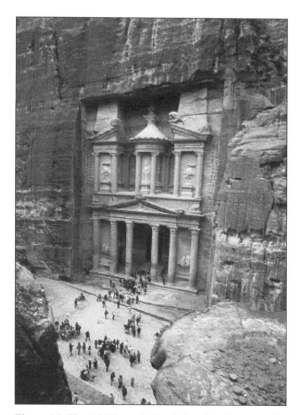

Figure 13. Graphs showing short-term fluctuations in relative humidity in the Al-Khazneh Chamber associated with the numbers of visitors in the chamber.

Figure 14. The Al-Khazneh Tomb façade when a tourist group gathers to enter the tomb chamber.

SUMMARY

The research that has been conducted to date in Petra, Jordan, includes examination of both intrinsic (i.e., lithology) and extrinsic (i.e., climate, human effects) influences on rock weathering, assists in establishing rates of deterioration that help forecast surface conditions, and indicates suitable methods that can be used to decrease natural and human-induced rates of deterioration. From 13 years of field research in Petra's Roman Theater, Anjar quarries, and Al-Khazneh tomb, we are able to better understand the complex dynamics of sandstone decay in Petra and in arid regions and to develop conservation strategies that will successfully protect this unique architecture from accelerated deterioration due to increasing regional and global tourism.

ACKNOWLEDGMENTS

This research was funded and supported by the National Science Foundation (NSF no. SES-9205055), U.S. Information Agency (USIA/USIS), the Jordanian-American Commission on Educational Exchange (JACE/CIES-Fulbright), the Petra National Trust, and the University of Arkansas King Fahd Center for Middle East and Islamic Studies, Fayetteville, Arkansas, USA. Field assistance was provided by Mick Frus, Mohammed Salem, and many of the B'dul residents.

REFERENCES CITED

Amoroso, G.G., and Fassina, V., 1983, Stone decay and conservation: Amsterdam, Elsevier, 453 p.

Blackwelder, E., 1929, Cavernous rock surfaces of the desert: American Journal of Science, v. 217, p. 393–399.

Blackwelder, E., 1933, The insolation hypothesis of rock weathering: American Journal of Science, v. 26, p. 97–113.

Bryan, K., 1922, Erosion and sedimentation in Papago country, Arizona: U.S. Geological Survey Bulletin 730-B, p. 19–90.

Bryan, K., 1928, Niches and other cavities in sandstone at Chaco Canyon, New Mexico: Zeitschrift für Geomorphologie, v. 3, p. 125–140.

Burton, R.F., 1879, The land of Midian: London, C. Kegan Paul and Company, 395 p.

Conca, J.L., and Rossman, G.R., 1982, Case hardening of sandstone: Geology, v. 10, p. 520–523, doi: 10.1130/0091-7613(1982)10<520:CHOS>2.0.CO;2.

Dragovich, D., 1986, Weathering rates of marble in urban environments, eastern Australia: Zeitschrift für Geomorphologie, v. 30, p. 203–214.

Emery, K.O., 1960, Weathering of the Great Pyramid: Journal of Sedimentary Petrology, v. 30, p. 140–143.

Fahey, B.D., and Dagresse, D.F., 1984, An experimental study of the effect of humidity and temperature variations on the granular disintegration of argillaceous carbonate rocks in cold climates: Arctic and Alpine Research, v. 16, p. 291–298.

Fahey, B.D., and LeFebure, T.H., 1988, The freeze-thaw weathering regime at a section of the Niagara Escarpment on the Bruce Peninsula, southern Ontario, Canada: Earth Surface Processes and Landforms, v. 13, p. 293–304.

Geikie, A., 1880, Rock weathering in Edinburgh churchyard: Proceedings of the Royal Society of Edinburgh, v. 10, p. 518–532.

Griggs, D.T., 1936, The factor of fatigue in rock exfoliation: Journal of Geology, v. 44, p. 781–796.

Hall, K., 1986, Rock moisture content in the field and laboratory and its relationship to mechanical weathering studies: Earth Surface Process and Landforms, v. 11, p. 131–142.

Herodotus, 450 BCE, The histories: New York, Penguin Books (reprinted 1978), 653 p.

Hoke, E., 1978, Investigation of weathering crusts on Salzburg stone monuments: Studies in Conservation, v. 23, p. 118–126.

Jackson, T.A., and Keller, W.D., 1970, A comparative study of the role of lichens and inorganic processes in the chemical weathering of recent Hawaiian lava flows: American Journal of Science, v. 269, p. 446–466.

Jenkins, K.A. and Smith, B.J., 1990, Daytime rock surface temperature variability and its implications for mechanical rock weathering, Tenerife, Canary Islands: CATENA, v. 17, p. 449–459, doi: 10.1016/0341-8162(90)90045-F.

Jones, D., Wilson, M.J., and Tait, J.M., 1980, The weathering of basalt by *Pertusaria corallina*: Lichenologist, v. 12, p. 277–289.

Kerr, A., Smith, B.J., Whalley, W.B., and McGreevy, J.P., 1984, Rock temperatures from S.E. Morocco and their significance for experimental rock-weathering studies: Geology, v. 12, p. 306–309, doi: 10.1130/0091-7613(1984)12<306:RTFSMA>2.0.CO;2.

Klein, M., 1984, Weathering rates of limestone tombstones measured at Haifa, Israel: Zeitschrift für Geomorphologie, v. 28, p. 105–111.

Matthias, G.F., 1967, Weathering rates of Portland arkose tombstones: Journal of Geological Education, v. 15, p. 140–144.

Meierding, T.C., 1981, Marble tombstone weathering rates: A transect of the United States: Physical Geography, v. 2, p. 1–18.

Mustoe, G.E., 1983, Cavernous weathering in the Capitol Reef Desert, Utah: Earth Surface Processes and Landforms, v. 8, p. 517–526.

Neil, D., 1989, Weathering rates of subaerially exposed marble in eastern Australia: Zeitschrift für Geomorphologie, v. 33, p. 464–473.

Paradise, T.R., 1995, Sandstone Weathering Thresholds in Petra, Jordan: Physical Geography, v. 16, p. 205–222.

Paradise, T.R., 1997, Disparate weathering from lichen overgrowth, Red Mountain, Arizona: Geografiska Annaler, v. 79, p. 177–184, doi: 10.1111/j.0435-3676.1997.00014.x.

Paradise, T.R., 1998, Limestone weathering and rate variability, Great Temple, Amman, Jordan: Physical Geography, v. 19, p. 133–146.

Paradise, T.R., 1999a, Deterioration of classical sandstone architecture in Petra, Jordan, *in* Bisheh, G., ed, Annual of the Department of Antiquities of Jordan (ADAJ): Amman, Jordan National Press, p. 353-368.

Paradise, T.R., 1999b, Environmental setting and stone weathering, *in* Joukowsky, M., ed., Petra's Southern Temple: Providence, Brown University, p. 63–87.

Paradise, T.R., 2000, Sandstone architectural deterioration in Petra, Jordan: Influences and rates: Proceedings from the International Congress on Culture and Monuments UNESCO/ICCROM, Venice, Italy, 19 p.

Paradise, T.R., 2002, Sandstone weathering and aspect in Petra, Jordan: Zeitscrift für Geomorphologie, v. 46, p. 1–17.

Paradise, T.R., and Yin, Z.Y., 1993, Weathering pit characteristics and topography, Stone Mountain, Georgia: Physical Geography, v. 14, p. 68–80.

Pfluger, F., 1995, Archaeo-geology in Petra, Jordan: Annual of the Department of Antiquities of Jordan, v. 39, p. 281–295.

Pliny the Elder (Gaius Plinius Secundus) (CE 50), Naturalis Historia (reprinted 1991): New York, Penguin, 399 p.

Pope, G.P., Meierding, T.C., and Paradise, T.R., 2002, Geomorphic approach to weathering studies in cultural resource management (CRM): Geomorphology, v. 47, p. 211–225, doi: 10.1016/S0169-555X(02)00098-3.

Rahn, P.H., 1971, Weathering of tombstone and its relationship to the elevation topography of New England: Journal of Geological Education, v. 19, p. 112–118.

Robinson, D.A., and Williams, R.B., 1992, Sandstone weathering in the High Atlas, Morocco: Zeitschrift für Geomorphologie, v. 36, p. 413–429.

Robinson, D.A., and Williams, R.B.G., 1989, Polygonal cracking of sandstone at Fontainebleau, France: Zeitschrift für Geomorphologie, v. 33, p. 59–72.

Roth, E.S., 1965, Temperature and water content as factors in desert weathering: Journal of Geology, v. 73, no. 3, p. 454–468.

Sancho, C., and Benito, G., 1990, Factors controlling tafoni weathering, Ebro Basin, Spain: Zeitschrift für Geomorphologie, v. 34, p. 165–177.

Shaer, M., 1997, The Nabataean Mortars in the Petra Area: Investigations of types and applications [M.A. Thesis]: Irbid, Jordan, Yarmouk University Institute of Archaeology and Anthropology.

Smith, B.J., 1977, Rock temperature measurements in northeast Sahara and their implications for rock weathering: CATENA, v. 4, p. 41–63, doi: 10.1016/0341-8162(77)90011-X.

Smith, B.J., and Magee, R.W., 1990, Granite weathering in an urban environment: An example from Rio de Janeiro: Singapore Journal of Tropical Geography, v. 11, no. 2, p. 143–153.

Smith, B.J., and McGreevy, J.P., 1988, Contour scaling of a sandstone by salt weathering under simulated hot desert conditions: Earth Surface Processes and Landforms, v. 13, p. 697–705.

Spencer, T., 1981, Microtopographic change on calcarenites, Grand Cayman Island, West Indies: Earth Surface Processes and Landforms, v. 6, p. 85–94.

Sperling, C.H., and Cooke, R.U., 1985, Laboratory simulation of rock weathering by salt crystallization and hydration processes in hot arid environments: Earth Surface Processes and Landforms, v. 10, p. 541–555.

Stephens, J.L., 1837, Incidents of Travel in Egypt, Arabia Petraea and the Holy Land (reprint edition 1996): Mineola, New York, Dover Publications, 473 p.

Strabo (CE 10), Geography (reprinted 1857): London, H.G. Bohn Publishers, 519 p.

Taylor, J., 2002, Petra and the Lost Kingdom of the Nabataeans: Harvard, Harvard University Press, 224 p.

Trudgill, S.T., 1989, Remeasurement of weathering rates, St. Paul's Cathedral, London: Earth Surface Processes and Landforms, v. 14, p. 175–196.

Winkler, E., 1965, Weathering rates as exemplified by Cleopatra's Needle, New York City: Journal of Geological Education, v. 13, p. 50–52.

Young, A.R., 1987, Salt as an agent in the development of cavernous weathering: Geology, v. 15, p. 962–966, doi: 10.1130/0091-7613(1987)15<962:SAAAIT>2.0.CO;2.

Manuscript Accepted by the Society 19 January 2005

Printed in the USA

Geological Society of America
Special Paper 390
2005

Characterization of swelling in clay-bearing stone

George W. Scherer
Inmaculada Jimenez Gonzalez
Princeton University, Department of Civil & Environmental Engineering,
Engineering Quad. E-319, Princeton, New Jersey 08544, USA

ABSTRACT

Many sedimentary rocks contain clays that cause differential swelling upon exposure to moisture, and the resulting internal stresses are blamed for the deterioration of buildings and monuments. To predict the likelihood of damage from this mechanism, it is necessary to characterize the magnitude of the swelling and the mechanical properties of the stone. Stones that swell also exhibit viscoelastic behavior, probably owing to sliding of the clay layers. In this paper we discuss the characterization of the relaxation and swelling behavior and the estimation of stresses resulting from swelling. A new method for measuring swelling is introduced, in which warping of a plate of stone is produced by wetting one side. This method is faster than the traditional direct measurement of swelling pressure, and it also yields information about the permeability and the influence of wetting on the elastic modulus. Sample results are presented for Portland Brownstone.

Keywords: Portland Brownstone, swelling clays, viscoelastic behavior, permeability, Young's modulus of elasticity, sedimentary rocks, warping, water absorption, mechanical properties, porosity, beam-bending method, stress, strain, sorptivity, crystallization pressures.

INTRODUCTION

Many sedimentary rocks contain clays that cause the stone to swell upon exposure to moisture (Dunn and Hudec, 1966), and the resulting deformation is thought to be responsible for deterioration of buildings, monuments, and sculptures (Delgado Rodrigues, 2001; Veniale et al., 2001). Examples of susceptible stone include the brownstone widely used in the northeastern United States (Jimenez Gonzalez et al., 2002), the molasse used in the cathedral of Lausanne, Switzerland (Félix, 1994), and sandstones used in some Egyptian sculpture (Charola et al., 1982; Rodriguez-Navarro et al., 1997). Figure 1 shows damage to the molasse in the cathedral in Lausanne. On the left side of the window (Fig. 1, top photo) there are large cracks indicating

separation of the surface layer from the underlying stone; on the right side (Fig. 1, bottom photo) the damage has progressed farther, and the surface layers have detached. Looking at the surface of the lower part of the column in Figure 1, it is evident that a relatively coherent surface layer >1 cm thick has detached along a circumferential plane. This is the sort of damage that would be expected if the surface expanded relative to the interior.

Although some clays exhibit enormous expansions, the greatest stress is exerted at small strains; the pressure required to suppress the expansion of clay typically drops exponentially as the volume increases (Macey, 1942; MacEwan and Wilson, 1980). However, only a small strain is required to produce stresses exceeding the tensile strength of stone, so differential swelling resulting from superficial wetting or drying of a stone surface can cause deterioration even for a stone whose swelling is on the order of 0.1%.

E-mail: scherer@princeton.edu

Scherer, G.W., and Jimenez Gonzalez, I., 2005, Characterization of swelling in clay-bearing stone, *in* Turkington, A.V., ed., Stone decay in the architectural environment: Geological Society of America Special Paper 390, p. 51–61, doi: 10.1130/2005.2390(06). For permission to copy, contact editing@geosociety.org.
©2005 Geological Society of America.

Figure 1. (Upper) Cracks in stone (molasse) on left side of window in Lausanne Cathedral; (Lower) damaged stone on right side of the same window.

The goal of this research is to assess the risk of damage from swelling. The problem is complicated by the fact that stones that swell are also viscoelastic, and the dynamic elastic modulus obtained from acoustic measurements is vastly different from the static modulus that is relevant for stress development (Tutuncu et al., 1998). This paper presents analyses of swelling experiments that are useful for evaluating the stresses produced by swelling. First, the standard "direct test" is considered, where a sample of stone is held at a fixed length in a rigid frame and then saturated with water. Normally, one simply measures the force exerted by the stone against the constraint, but it is also possible to extract information about the kinetics of water movement and viscoelastic relaxation of the stone. This measurement has certain disadvantages, so a new method for evaluating swelling stress is introduced, which is to observe the warping of a thin plate of stone when water is put in contact with one side. These methods have been used in a detailed study of the behavior of Portland Brownstone and several other stones. The purpose of this paper is to examine the measurement methods in detail.

This study of swelling indicates that some stones may indeed suffer destructive stresses as a result of the differential strains produced during wetting. However, it is important to recognize

that stones containing clays necessarily contain a fraction of very small pores, and these are the pores most likely to generate high crystallization pressure (Scherer, 1999). Consequently, such stones are inherently susceptible to damage from crystallization of ice and salt (McGreevy and Smith, 1984). The fact that these two mechanisms of damage are likely to act in concert makes it particularly important to evaluate their relative importance so that appropriate conservation measures can be chosen.

PORTLAND BROWNSTONE

To illustrate the application of the swelling characterization experiments, some results of experiments on Portland Brownstone are used here to offer a brief description of the materials and procedures.

Structure

Plates ($25 \times 25 \times 5$ cm) of Portland Brownstone were purchased from Pasvalco Company of Closter, New Jersey, and samples were prepared by cutting with a diamond saw or core drill (2 cm inner diameter). X-ray analysis performed on one of our samples (T.

Fuellmann, 2002, personal commun.) indicated that it consisted primarily of grains of quartz and anorthite in a matrix containing montmorillonite, kaolinite, and illite. The pore structure was characterized by nitrogen sorption using a Micromeritics ASAP 2010 and by mercury porosimetry using a Micromeritics 9410. The pore diameter distribution was found to be bimodal (Jimenez Gonzalez et al., 2002), with about half the volume around a peak near 10 μm and the rest near a peak at 0.1 μm. The latter peak was the only one detected by nitrogen sorption. The porosity of the stone, obtained by vacuum impregnation with water, is ~8.7%.

Free Swelling Strain

To measure the free expansion of the stone, rectangular parallelepipeds or core-drilled cylinders were placed into the differential mechanical analyzer (DMA, Perkin-Elmer). The maximum sample height in this instrument is limited to ~18 mm. A bath of water was raised around the sample, and the linear expansion was measured as a function of time. The swelling was found to be $4.5 \pm 1.0 \times 10^{-4}$ perpendicular to the bedding.

Sorptivity

Sorptivity was measured by suspending samples of stone from an electronic balance so that the bottom of the stone touched a large pool of water and recording the output from the balance continuously; the sides of the sample were coated with grease to prevent evaporation. The height of rise in the stone, h, is calculated from the weight gain, Δm, using

$$h = \frac{\Delta m}{\phi \rho_L A},$$ (1)

where A is the area of the sample in contact with the liquid, ϕ is the porosity of the stone ($\phi = 0.087$ for the Portland Brownstone used in our study), and $\rho_L = 1.0$ g/cm^3 is the density of the liquid. The sorptivity, S, is then defined by

$$h = S \sqrt{t},$$ (2)

where t is the time in seconds. The sorptivity of the Portland Brownstone was found to be $S_r \approx 0.012$ cm/s$^{1/2}$ parallel to the bedding and $S_z = 0.0079$ cm/s$^{1/2}$ perpendicular to the bedding.

For comparison to other measurements, it is useful to note that the sorptivity is related to the permeability, k. The water moves into the stone according to Darcy's law (Happel and Brenner, 1986),

$$J = -\frac{k}{\eta_L} \nabla p,$$ (3)

where J is the flux, η_L is the viscosity of the liquid, and ∇p is the pressure gradient. In the sorptivity experiment, the gradient is

$$\nabla p = \frac{p_c}{h} = \frac{2\gamma}{h r_m},$$ (4)

where ∇p is the capillary pressure ($\nabla p < 0$), γ is the liquid-vapor interfacial energy, and r_m is the radius of the meniscus; if the contact angle of the liquid on the pore walls is zero, then r_m equals the pore radius, r_p. Recognizing that the flux in the sorptivity experiment is $\phi \, dh/dt$, the sorptivity is found to be

$$S = \sqrt{\frac{-2k \, p_c}{\phi \eta_L}} \approx \sqrt{\frac{4k \, \gamma}{\phi \eta_L r_p}}.$$ (5)

Elastic Modulus Measurements

The dynamic elastic modulus was calculated from the acoustic velocity at 54 KHz, measured using a commercial instrument (PUNDIT, CNS Farnell, London, UK) on stones of various sizes, wet and dry. The size of the sample is significant, because this method yields Young's modulus on very slender bodies and the longitudinal modulus on very large ones (Kolsky, 1963). More important is the effect of strain on the modulus (Tutuncu et al., 1998), because the acoustic method imposes a very small strain, and the modulus decreases as the strain increases. The frequency also has a large affect on the modulus, particularly when the sample is wet. The dynamic modulus of the dry stone was ~25 GPa and rose to 35 GPa in the wet stone. However, the static modulus (obtained from the slope of the 3-point bending curves) was only ~9.1 GPa in the dry stone and decreased to ~4.1 GPa in samples saturated with water. Similar enormous discrepancies between acoustic and static moduli have been reported (Tutuncu et al., 1998). The effect is attributed to the movement of water from grain contacts, which provides compliance at low frequencies of loading, but which cannot occur at high frequencies. Since we are interested in stresses generated over a long period of time under substantial strains, only the static modulus data are relevant.

When a plate of saturated stone is subjected to bending, the concave side of the sample is compressed and the convex side is stretched, so a pressure gradient is created in the pore liquid. The liquid flows until atmospheric pressure is restored, and the force required to sustain a fixed deflection decreases with time; this process is called hydrodynamic relaxation, because it results from flow of the pore liquid. If the solid phase in the sample is viscoelastic, then further relaxation of the force will occur. This phenomenon has been used to measure the permeability of porous bodies, including gels (Scherer, 1992), porous glass (Vichit-Vadakan and Scherer, 2000), and cement paste (Vichit-Vadakan and Scherer, 2002). For this measurement, an apparatus (Vichit-Vadakan and Scherer, 2000) with a rigid frame holding a stepper motor was used. The pushrod is driven downward by the motor; a load cell is incorporated into the pushrod to measure the applied force, and the pushrod passes through a linear variable differential transformer (LVDT) that measures the displacement. The sample was supported at one end by a roller and at the other end by a ball, so that the plate was not subjected to twisting, even if it was not perfectly flat. The sample was immersed in liquid, so that the pore pressure was fixed at atmospheric pressure at the surfaces and liquid could freely exchange between the pores and the bath. Figure 2 shows the load relaxation of a plate of Portland Brownstone after application of a deflection of 136 µm. The fit of the theoretical curve (Scherer, 2000a) to the data is seen to be very good; it consists of the product of the hydrodynamic and visco-elastic relaxation curves, which are shown separately. The param-eters of the fit include the elastic modulus of the saturated plate, $E = 3.9$ GPa, and the permeability, $k = 0.15$ nm². Substituting this value into Equation (5), we find that $r_p = 40$ nm, which is in very good agreement with the mean size measured by nitrogen sorp-tion; this indicates that the smaller peak of the pore size distri-bution provides the driving force for capillary rise. The viscoelastic relaxation is approximated by a stretched exponential function,

$$\frac{\sigma_z(t)}{\sigma_z(0)} \equiv \Psi(t) = \exp\left[-\left(\frac{t}{\tau_{VE}}\right)^\beta\right], \qquad (6)$$

with $\tau_{VE} \approx 2 \times 10^{10}$ s and $\beta \approx 0.12$. Only a small part of the relax-ation is observed, so the fitting parameters cannot be taken too seriously (i.e., the relaxation process would probably not take 10^5 yr), but they do accurately represent the first 15% of the relaxation, which takes ~3 h.

Swelling Measurements

Direct measurements of swelling were performed on a sam-ple of Portland Brownstone ~5 cm high and 3.17 cm square, with a circular hole 2.18 cm in diameter. The arrangement, shown in Figure 3, allows water to invade the sample from the inner and outer surfaces simultaneously, while the height of the stone is fixed by the platens of the testing machine.

An alternative method for measuring expansion, discussed in detail in a later section, is to induce warping in a thin plate of stone by wetting only one side. For this purpose, a plate of Port-land Brownstone 0.38 cm thick and 1.0 cm wide was placed on supports 9.3 cm apart. A bead of vacuum grease was run around the upper edge (to act as a dam); water was poured onto the top surface, and the deflection of the plate was measured with an optical probe (MTI 1000, MTI Instruments, Inc.) mounted below the plate. A small piece of reflective foil was glued to the bottom to improve the sensitivity of the probe.

THEORETICAL BACKGROUND

Direct Measurement of Swelling Pressure

The conventional method (e.g., Madsen and Müller-Von-moos, 1989) for evaluating the swelling pressure is to place a sample of stone into a rigid frame and then saturate it with water. As the water invades the pores of the stone, the stone swells and

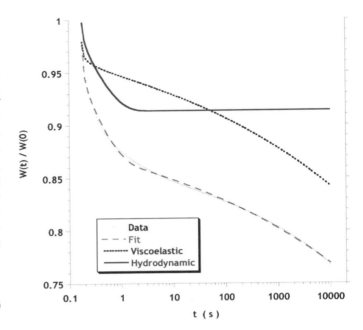

Figure 2. Relaxation of applied force at constant deflection for a plate of Portland Brownstone saturated with pure water. The dashed line is the fit of the data to the theoretical curve (Scherer, 2000a). The hydro-dynamic relaxation results from equilibration of the pressure in the pore liquid, which is complete in ~2 seconds. The viscoelastic curve represents the stress relaxation produced by flow of the solid phase; the curve is given by Equation (6).

exerts pressure on the confining frame. The objective of this mea-surement is usually to determine the maximum pressure exerted by the stone, but the kinetics of the increase provide information about the sorptivity of the stone. In this section, an elastic analy-sis of the stress is presented, then the kinetics of stress develop-ment is considered by examining the rate of penetration of the water into the stone, and the importance of viscoelastic effects is considered, given that the duration of the experiment is several hours so that substantial relaxation can occur.

The sedimentary rocks that contain clays often show strong bedding, so it should be expected that in general the stone is transversely isotropic; i.e., the properties are uniform parallel to the bedding, but different in the perpendicular direction. Let us adopt a cylindrical coordinate system with the z axis perpendicu-lar to the bedding, and the $r - \theta$ plane parallel to the bedding. The elastic constitutive equations are

$$\varepsilon_r = \frac{du}{dr} = \varepsilon_{fr} + \frac{\sigma_r}{E_r} - \frac{v_r\sigma_\theta}{E_r} - \frac{v_z\sigma_z}{E_z} \qquad (7)$$

$$\varepsilon_\theta = \frac{u}{r} = \varepsilon_{fr} + \frac{\sigma_\theta}{E_r} - \frac{v_r\sigma_r}{E_r} - \frac{v_z\sigma_z}{E_z} \qquad (8)$$

$$\varepsilon_z = \varepsilon_{fz} + \frac{\sigma_z}{E_z} - \frac{v_z\sigma_r}{E_z} - \frac{v_z\sigma_\theta}{E_z} \qquad (9)$$

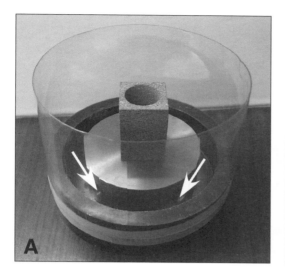 

Figure 3. Setup for direct swelling measurement. (A) Hollow sample of stone rests on stainless steel platform with holes (indicated by arrows) that allow water to flow up through hole centered under sample so that water penetrates from the inner and outer surfaces of the sample. (B) Sample held to fixed length by platens of testing machine. Once the sample is arranged as in (B), water is quickly poured into the dish until the sample is submerged.

where ε_j ($j = r,\theta,z$) are the strains, σ_j are the stresses, E_j and v_j are Young's modulus and Poisson's ratio, respectively, and u is the radial displacement. The free strain produced by swelling of the clay is ε_{fr} in the plane and ε_{fz} perpendicular to the bedding. As noted in the previous section, the latter quantity was found to be $\varepsilon_{fz} = 4.5 \pm 1.0 \times 10^{-4}$ for the Portland Brownstone used in this study. For a transversely isotropic material, Darcy's law becomes (Bear, 1972):

$$J_j = -\frac{k_j}{\eta_L} \nabla p \left(j = r, \theta, z \right), \qquad (10)$$

where $k_r = k_\theta$. In the warping experiment described above, the bedding was perpendicular to the long axis of the plate, so the permeability obtained was k_r. For an isotropic material, Equations (7–10) still apply, with $E_r = E_z$, $v_r = v_z$, and $k_r = k_z$.

The geometry of the direct stress measurement is shown in Figure 4. The bedding is horizontal and the water invades from the inner and outer surfaces of the sample, which is in the form of a hollow cylinder. The inner wet region (Region 1) extends from $r = a$ to $r = r_1$, and the outer wet region (Region 2) extends inward from $r = b$ to $r = r_2$; the dry region lies between r_1 and r_2. In the dry region, the free swelling strain is zero and in the wet region it is ε_{fz}. Force balance in the cylinder requires:

$$\sigma_\theta = \frac{d}{dr} \left(r\sigma_r \right). \qquad (11)$$

Substituting Equations (7) and (8) into Equation (11) shows that the radial displacement must have the form

$$u = c_1/r + c_2 r, \qquad (12)$$

where c_1 and c_2 are constants to be determined; there are two such constants for each of the three regions. The boundary conditions required to find those constants are obtained by requiring that the radial stresses be zero at the exposed surfaces ($r = a$ and b), and that the radial stresses and displacements be continuous at r_1 and r_2. Finally, the axial strains are required to be zero in all regions, because of the constraint imposed by the testing machine. Satisfying these equations leads to expressions for the axial stress in each region. The total force, F, on the testing machine is the sum of the forces from each region:

$$F = \left(r_1^2 - a^2 \right)\sigma_{z1} + \left(r_2^2 - r_1^2 \right)\sigma_{zd} + \left(b^2 - r_2^2 \right)\sigma_{z2}. \qquad (13)$$

The complete expression for the force for a transversely isotropic body is very complicated (it would take several pages to write out the constants), but a drastic simplification is obtained if we make the reasonable assumption that the softening produced by exposure to water is the same in the radial and axial directions:

$$\frac{E_{rw}}{E_{rd}} \approx \frac{E_{zw}}{E_{zd}} \equiv f, \qquad (14)$$

and that the value of Poisson's ratio is roughly the same in both directions ($v_r = v_z$) and is not significantly affected by wetting. With these approximations, Equation (13) reduces to

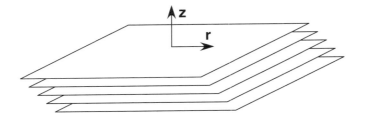

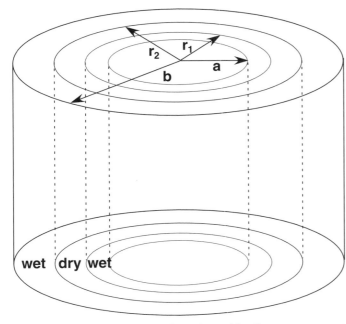

Figure 4. Schematic illustration of sample used for direct measurement of swelling pressure. The bedding is perpendicular to the axis of the cylinder as indicated at the top of the figure. Liquid invades through the inner and outer surfaces ($r = a$ and b), and the edges of the advancing fronts from the inner and outer surfaces are r_1 and r_2, respectively.

and the transport obeys Darcy's law. To solve for the rate of invasion, the continuity equation is solved for flow within the wet region (Scherer, 2000b):

$$\left(1 - \frac{K_p}{K_S}\right)\dot{\varepsilon} + \dot{p}\left(\frac{1 - K_p/K_S - \phi}{K_S} + \frac{\phi}{K_L}\right) = \nabla \cdot J_r, \quad (17)$$

where $\dot{\varepsilon}$ is the volumetric strain, ϕ is the volume fraction of porosity, K_L is the bulk modulus of the liquid, and liquid, K_S is the bulk modulus of the solid phase and K_p is the bulk modulus of the porous body. The maximum value for p is the capillary pressure, p_c, which is very small compared to the elastic moduli, so those terms can be neglected; similarly, the volumetric expansion is negligible (<0.1% linearly), so the equation reduces to

$$\nabla \cdot J_r = -\frac{k_r}{\eta_L}\nabla^2 p = 0. \quad (18)$$

Then, Equation (18) is solved subject to the following boundary conditions: the pore pressure at the free surfaces is zero, $p_1(a) = p_2(b) = 0$, and is equal to the capillary pressure, p_c, at the leading edge of the advancing liquid front, $p_1(r_1,t) = p_2(r_2,t) = p_c$. In region 1, the pressure is found to be

$$p_1(r,t) = \frac{p_c \ln[r/a]}{\ln[r_1(t)/a]}, \quad (19)$$

and in region 2,

$$p_2(r,t) = \frac{p_c \ln[b/r]}{\ln[b/r_2(t)]}. \quad (20)$$

The volume of water crossing the boundary of region 1 is given by

$$2\pi r_1 \frac{dr_1}{dt}\phi = 2\pi r_1 J_r(r_1,t), \quad (21)$$

which leads to

$$\int_a^{r_1} r_1 \ln[r_1/a]\, dr_1 = -\int_0^t \frac{k_r p_c}{\eta_L \phi}dt \quad (22)$$

and

$$\left(\frac{r_1^2}{2}\right)\ln\left[\frac{r_1}{a}\right] - \frac{r_1^2 - a^2}{4} = -\left(\frac{k_r p_c}{\eta_L \phi}\right)t. \quad (23)$$

Using Equation (5), we can define a dimensionless time:

$$\theta = \left(\frac{S_r}{a}\right)^2 t, \quad (24)$$

$$F = -\left(b^2 - r_2^2 + r_1^2 - a^2\right)E_{zw}\varepsilon_{fz}, \quad (15)$$

so the swelling stress is

$$\sigma_s = \frac{F}{b^2 - a^2} = -\left(\frac{b^2 - r_2^2 + r_1^2 - a^2}{b^2 - a^2}\right)E_{zw}\varepsilon_{fz}. \quad (16)$$

Once saturation is complete ($r_1 = r_2$), given the values previously cited for E_{zw} and ε_{fz}, the direct stress measurement is expected to yield $\sigma_s = -E_{zw}\varepsilon_{fz} \approx 1.4 - 2.3$ MPa for Portland Brownstone. The next problem is to find the rate of penetration of the water into the sample.

The water invades the sample just as in a sorptivity experiment, where the pressure gradient results from capillary pressure and Equation (23) is rewritten as

$$\frac{1}{2}\left[1 - \frac{r_1^2}{a^2} + \frac{r_1^2}{a^2}\ln\left(\frac{r_1^2}{a^2}\right)\right] = \theta. \tag{25}$$

Similarly, considering the flux from region 2,

$$\int_b^{r_2} r_2 \ln\left[b/r_2\right] dr_2 = \int_0^t \frac{k_r p_c}{\eta_L \phi} dt \tag{26}$$

so,

$$\frac{1}{2}\left[1 - \frac{r_2^2}{b^2} + \left(\frac{r_2^2}{b^2}\right)\ln\left(\frac{r_2^2}{b^2}\right)\right] = \left(\frac{S_r}{b}\right)^2 t = c^2\theta, \tag{27}$$

where $c = a/b$. The sample becomes saturated when $r_1 = r_2$. Equating Equations (23) and (28), the depth at which the advancing fronts meet is

$$r_1^{sat} = r_2^{sat} = a\sqrt{\frac{1/c^2 - 1}{\ln\left(1/c^2\right)}}. \tag{28}$$

Substituting those dimensions into Equations (25) or (27), the time at which saturation is achieved is

$$\theta_{sat} = \left(\frac{S_r}{a}\right)^2 t_{sat} = \frac{1}{2} - \frac{1 - c^2}{2c^2\log\left(1/c^2\right)}\left[\log\left(\frac{c^2\log\left(1/c^2\right)}{1 - c^2}\right) + 1\right]. \tag{29}$$

For $0.5 \leq a/b \leq 1$, θ_{sat} is approximated with less than 2% error by

$$\theta_{sat} \approx \frac{1}{4}\left(\frac{1}{c} - 1\right)^2. \tag{30}$$

The measured stress depends on the volume fraction of saturation, v, which is defined as

$$v = \frac{b^2 - r_2^2 + r_1^2 - a^2}{b^2 - a^2}. \tag{31}$$

Using Equations (25) and (27) to evaluate Equation (31), when $a/b \geq 0.5$, the volume fraction saturated is given to a very high accuracy by

$$v \approx \sqrt{\frac{\theta}{\theta_{sat}}} = \sqrt{\frac{t}{t_{sat}}}. \tag{32}$$

Therefore, Equation (16) can be written as a simple function of time:

$$\sigma_z = -vE_{zw}\varepsilon_{fz} = -E_{zw}\varepsilon_{fz}\sqrt{\frac{t}{t_{sat}}}. \tag{33}$$

The advantage of this experiment is that it provides a direct measurement of the quantity $E_{zw}\varepsilon_{fz}$ (obtained from the final stress value); moreover, the sorptivity can be found from the time required to achieve saturation, since Equations (29) and (30) indicate that

$$t_{sat} \approx \frac{1}{4}\left(\frac{b - a}{S}\right)^2. \tag{34}$$

However, this analysis is only valid when the bedding is horizontal; otherwise, the sample is not axisymmetric, and the analysis is quite complicated. Moreover, the time required to achieve saturation is typically hours or days, during which time there will be substantial viscoelastic relaxation. Consequently, the elastic analysis presented here is not adequate to describe the kinetics.

A viscoelastic (VE) analysis can be obtained quite simply for a planar geometry, and since Equation (32) indicates that the kinetics for the approach to saturation are nearly identical for a cylinder and a plate, the planar solution should be a good approximation to the present case. First, an elastic solution for the plate is obtained, then the VE solution follows directly. Suppose that water invades the opposite faces of a plate of stone whose thickness is L, so that each side is saturated to a depth L_w and there is a dry central region of width L_d, where $L = L_d + 2L_w$. If the water is advancing parallel to the x axis and the expansion is constrained in the z direction, then the strain $\varepsilon_z = 0$ and the stresses in the unconstrained directions are $\sigma_x = \sigma_y = 0$. The constitutive equation in the z direction is given by Equation (9):

$$\varepsilon_z = \varepsilon_{fz} + \frac{\sigma_{zw}}{E_{zw}} = 0, \tag{35}$$

where σ_{zw} and E_{zw} are the stress and Young's modulus in the wet region, and ε_{fz} is the free swelling strain. In the dry interior, there is no free strain, so

$$\varepsilon_z = \frac{\sigma_{zd}}{E_{zd}} = 0. \tag{36}$$

Thus, there is no stress in the dry region, and the total stress exerted by the swelling plate is

$$\sigma_s = \left(\frac{2L_w}{L}\right)\sigma_{zw} = -\left(\frac{2L_w}{L}\right)E_{zw}\varepsilon_{fz}. \tag{37}$$

From Equation (2), $L_w = S_r\sqrt{t}$, so the elastic solution for the swelling stress is

$$\sigma_s^E = -2\left(\frac{S_r\sqrt{t}}{L}\right)E_{zw}\varepsilon_{fz}. \tag{38}$$

If the wet stone is viscoelastic, then Equation (6) indicates that E_{zw} must be replaced by $E_{zw}\psi(t)$. The water reaches a depth x at time $t' = x^2/S_r^2$, and that is the time at which relaxation begins, so the stress at any point is

$$\sigma_{zw}^{VE}(x,t) = -E_{zw}\varepsilon_{fz}\Psi\left(t - x^2/S_r^2\right). \quad (39)$$

Thus, for a VE material, Equation (38) is replaced by

$$\sigma_s^{VE} = \left(\frac{2}{L}\right)\int_0^{L_w}\sigma_{zw}dx = -\left(\frac{2}{L}\right)E_{zw}\varepsilon_{fx}\int_0^{L_w}\Psi\left(t - x^2/S_r^2\right)dx. \quad (40)$$

The integral can be written in terms of time alone, since $x = S_r\sqrt{t}$:

$$\sigma_s^{VE} = -\left(\frac{S_r}{L}\right)E_{zw}\varepsilon_{fz}\int_0^t\Psi\left(t - t'\right)\frac{dt'}{\sqrt{t'}}, \; t \leq t_{sat}. \quad (41)$$

When the material is elastic, $\psi(t) = 1$, and Equation (41) reduces to Equation (38). Equation (41) applies until the plate becomes saturated, at time $t_{sat} = L^2/(4S_r^2)$; thereafter,

$$\sigma_s^{VE} = -\left(\frac{S_r}{L}\right)E_{zw}\varepsilon_{fz}\int_0^{t_{sat}}\Psi\left(t - t'\right)\frac{dt'}{\sqrt{t'}}, \; t \leq t_{sat}. \quad (42)$$

Unfortunately, if the relaxation of the stone obeys the stretched exponential function in Equation (6), Equation (41) cannot be integrated in closed form.

If the VE relaxation is slow compared to the duration of the swelling experiment, then the elastic analysis is sufficient, and one can obtain both the maximum swelling stress and the sorptivity of the sample from this measurement. If relaxation is fast, then the measurement can be used to determine Ψ by fitting the measured stress to Equation (42). The simplest way to determine whether VE relaxation is sufficient is to continue the experiment well beyond the time of saturation to see whether the stress passes through a maximum and relaxes again. This type of experiment was performed on the sample in Figure 3; the result is shown in Figure 5. The calculated curve was obtained from Equations (41) and (42), using $\psi(t)$ from Figure 2, and assuming $S = 0.012$ cm/s$^{1/2}$; the peak stress value corresponds to $E_{zw}\varepsilon_{fz} = 0.74$ MPa. If $E_{zw} = 3.9$ GPa, as found from the fit in Figure 2, then $\varepsilon_{fz} = 1.4 \times 10^{-4}$, which is unusually small for this stone. The reproducibility of this type of experiment was poor and always yielded smaller than expected stresses. Part of the problem is that the force is at the very low end of the range of the load cell on the testing machine; moreover, the total expected free swelling is ~20 μm, so a very small movement of the platen would substantially affect the measured stress.

Although the direct stress measurement is rich in information, it is slow, difficult, and requires expensive equipment. Ordinary mechanical testing machines cannot necessarily be trusted to hold the sample dimensions fixed within a few microns, so specialized equipment should be built. Moreover, the experiment uses a relatively large amount of stone (especially if a mechanical testing machine with a large load cell is used), which may be problematic if the stone is to be taken from a monument. Therefore, it is desirable to have a test of swelling stress that is relatively fast and uses a minimal amount of stone. The warping method described in the next section satisfies those requirements.

Warping Method

A novel method for characterizing the swelling stress is presented here, based on the following principle (see Fig. 6): if one side of a thin plate of swelling stone is exposed to water, that side will swell and cause the plate to warp. The displacement of the plate is large, if the ratio of length to thickness is large, so the effect is easily measured.

An example of a warping experiment, performed on a plate of Portland Brownstone 0.38 cm thick, is shown in Figure 7. The upward deflection peaks at 155 μm ~100 s after the water is poured onto the plate; as the water proceeds through the plate, it becomes uniformly dilated, so the warping is eliminated. If the elastic properties of the wet and dry regions were equal, then the maximum deflection would occur when the water had penetrated halfway through the plate; however, the wet stone is less stiff, so the maximum is shifted. An elastic analysis of this problem is presented here; a viscoelastic treatment is not necessary, because the duration of the experiment is so short.

If, as illustrated in Figure 8, a long slender beam rests on supports with a span of L, and the plate acquires a radius of curvature R, then the deflection out of plane is Δ. If $\Delta \ll R$, then the geometry indicates that

$$\Delta = \frac{L^2}{8R}. \quad (43)$$

In the present case, the radius of curvature results from the difference in free strain between the wet part (strain = ε_{fz} or ε_{fr}, according to the orientation of the sample) and the dry part (strain = zero). The warping of a composite beam was analyzed by Timoshenko (1925; see Scherer, 1992), who found that:

$$\frac{1}{R} = \frac{K_R\Delta\varepsilon_f}{h}, \quad (44)$$

where h is the thickness of the plate, $\Delta\varepsilon_f = \varepsilon_{fz}$ or ε_{fr}, and

$$K_R = \frac{6m(1+m)^2 f}{m^4 f^2 + 4m^3 f + 6m^2 f + 4mf + 1}, \quad (45)$$

where $m = h_w/h$ and $f = E_{zw}/E_{zd}$ or E_{rw}/E_{rd}, according to the orientation of the plate. According to Equation (2),

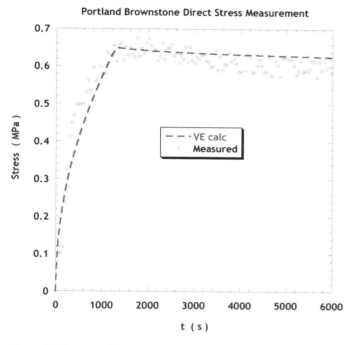

Portland Brownstone Direct Stress Measurement

Stress (MPa) vs *t (s)*

- - - VE calc
○ Measured

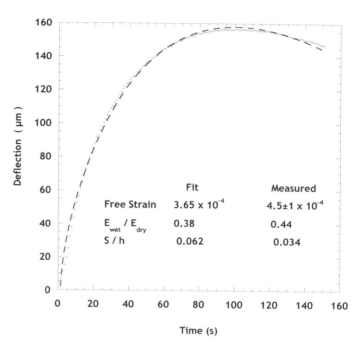

Deflection (μm) vs *Time (s)*

	Fit	Measured
Free Strain	3.65×10^{-4}	$4.5 \pm 1 \times 10^{-4}$
E_{wet} / E_{dry}	0.38	0.44
S / h	0.062	0.034

Figure 5. Direct swelling stress measurement for Portland Brownstone sample confined between platens of testing machine. The dashed curve was calculated using Equations (41) and (42), using $\psi(t)$ from Figure 2 and assuming $S = 0.013$ cm/s$^{1/2}$; the peak stress value corresponds to $E_{zw}\,\varepsilon_{fz} = 0.74$ MPa, which is smaller than expected (see text).

Figure 7. Measured deflection (symbols) and fit (dashed curve) to Equation (47). The fit is applied only to the data shown. At longer times, there is a discrepancy, which is shown in Figure 9.

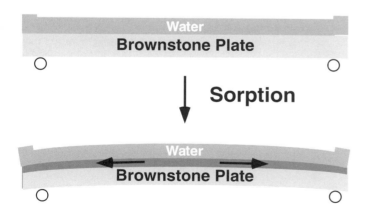

Figure 6. Schematic illustration of warping experiment. A plate of stone is supported at each end, and a dam of vacuum grease is applied around the upper edge. When water is poured on top, the upper surface of the stone expands and the plate warps; the upward deflection is measured with an optical probe (not shown) focused on the underside of the plate.

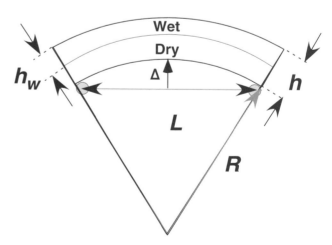

Figure 8. Schematic of warping showing radius of curvature, R, span between supports, L, and deflection, Δ; h is thickness of plate and h_w is depth of saturated zone.

$$m = \frac{h_w}{h} = \frac{S\sqrt{t}}{h} \equiv \sqrt{\theta}, \tag{46}$$

where the latter definition is analogous to Equation (24). If the bedding is perpendicular to the long axis of the plate, then $\Delta\varepsilon_f = \varepsilon_{fz}$, $S = S_r$, and $f = E_{zw}/E_{zd}$; if the bedding is parallel to the large surface of the plate, then $\Delta\varepsilon_f = \varepsilon_{fr}$, $S = S_z$, and $f = E_{rw}/E_{rd}$. With these abbreviations, Equation (43) can be written as

$$\Delta = \Delta_0 \left[\frac{4f\left(\theta^{1/2} - \theta\right)}{1 - 4\left(1 - f\right)\theta^{1/2} + 6\left(1 - f\right)\theta - 4\left(1 - f\right)\theta^{3/2} + \left(1 - f\right)^2 \theta^2} \right], \tag{47}$$

where,

$$\Delta_0 = \frac{3L^2\Delta\varepsilon_f}{16h}. \tag{48}$$

By setting the derivative of Δ with respect to θ equal to zero, the maximum in Δ is found to be equal to Δ_0, and the maximum occurs at reduced time:

$$\theta_{max} = \frac{1}{\left(1 + \sqrt{f}\right)^2} \tag{49}$$

or elapsed time:

$$t_{max} = \left(\frac{h}{S\left(1 + \sqrt{f}\right)} \right). \tag{50}$$

Thus, the free swelling strain of the stone can be found from the maximum deflection by using Equation (48). The quantities f and S can be obtained by fitting Equation (47) to the measured deflection, as in Figure 7. Alternatively, if the deflection is normalized by its maximum value (Δ_0) and the time is normalized by t_{max}, the resulting function depends only on f. Then it is only necessary to perform a one-parameter fit to

$$\frac{\Delta}{\Delta_0} = \left(4f\left(1 + \sqrt{f} - y\right)y\right) \Big/ \left(\left(1 + \sqrt{f}\right)^2 - 4\left(1 + \sqrt{f} - f - f^{3/2}\right)y\right.$$

$$\left. + 6\left(1 - f\right)y^2 - 4\left(1 - \sqrt{f}\right)y^3 + \left(1 - \sqrt{f}\right)^2 y^4\right) \tag{51}$$

where $y = \sqrt{t/t_{max}}$. Once f is obtained from this fit, the sorptivity can be calculated from t_{max} using Equation (50). An example of a fit to normalized data is shown in Figure 9.

DISCUSSION

The fit to the warping deflection in Figure 7 yields parameters that are in good agreement with independent measurements.

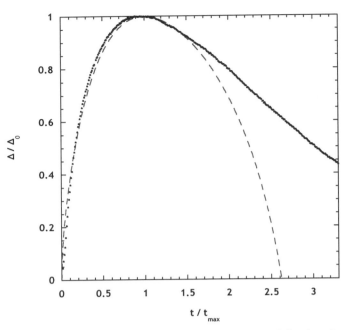

Figure 9. Data from Figure 7 normalized by the peak deflection, Δ_0, and the time of the maximum, t_{max}. The fit (dashed curve) is extrapolated to show the discrepancy between the measurement (symbols) and the fit at long times.

The plate was oriented with the bedding vertical (i.e., perpendicular to the long axis of the plate), so the free swelling strain is ε_{fz}. The fit indicates that $\varepsilon_{fz} = 3.65 \times 10^{-4}$, whereas independent measurements of free swelling described earlier give $4.5 \pm 1.0 \times 10^{-4}$. The fit yields $f = 0.38$, whereas the static modulus measurements described gave $f \approx 0.45$. The sorptivity found from the fit is almost double the average of the direct sorptivity measurements. It is not possible to obtain a good fit by reducing S_r and adjusting the other parameters, because each variable affects the shape of the curve in a different way. This particular sample might have been exceptionally permeable. Alternatively, the high value of sorptivity might reflect the relatively rapid invasion of large pores that do not percolate through the body. In a materials with a wide distribution of pore size, such as Portland Brownstone (Jimenez Gonzales et al., 2002), the water is expected to move some distance into the stone before a steady-state is reached, at which point the rate of advance is controlled by an average pore size (viz., the breakthrough radius; see Katz and Thompson, 1987). Since the warping sample is thin, the water might penetrate a significant fraction of the thickness before the steady-state sorptivity applies.

The fit was only applied to the range of data shown in Figure 7. At longer times, the relaxation of the plate is always slower than predicted, as shown in Figure 9. The effect is too large to result from viscoelastic relaxation and too persistent to result from hydrodynamic relaxation. It is suspected that the swelling strain is not achieved instantly, so the wet stone continues to expand for some time after the advancing water front passes by. This would cause the upper surface to continue to expand relative to the lower part (which has been in contact with the water for a shorter time), so the duration of the deflection would not be entirely controlled by the sorptivity but would last longer. This would also account for the fact that the warping method usually yields a somewhat smaller free strain than the DMA measurements. This point is currently under investigation.

CONCLUSIONS

To predict whether swelling will cause damage to stone, it is necessary to know the stresses that develop and the strength of the stone. Analysis of the stresses is complicated by viscoelastic relaxation, and it can be expected that the strength of the stone will also be dependent on the rate and duration of loading. It has been demonstrated (Girardet, 2003, personal commun.) that when molasse is loaded slowly over the course of 24 h, it breaks at ~1/3 of the compressive strength measured at standard rates. In this work, methods for measuring the properties that must be known in order to calculate the stresses are discussed. The beam-bending method yields the static modulus and stress relaxation kinetics; if the measurement is performed on a saturated sample, the permeability can also be obtained from the rate of hydrodynamic relaxation. The swelling strain can be measured with a dilatometer (such as a DMA), but we find it convenient to use a simple warping experiment, which yields the swelling strain, sorptivity, and the ratio of wet to dry modulus, all in a matter of minutes. This technique is particularly useful for screening stones for swelling and for evaluating chemical treatments (such as the surfactants proposed by Snethlage and Wendler [1991]) for suppression of swelling.

ACKNOWLEDGMENTS

Inmaculada Jimenez Gonzalez thanks the Kress Foundation for their financial support of this research.

REFERENCES CITED

Bear, J., 1972, Dynamics of fluids in porous media: American Elsevier, New York, p. 124, 136.

Charola, A.E., Wheeler, G.E., and Koestler, R.J., 1982, Treatment of the Abydos reliefs: Preliminary investigations, *in* Gauri, K.L., and Gwinn, J.A., eds., Fourth International Congress on Deterioration and Preservation of Stone Objects: Louisville, Kentucky, University of Louisville, p. 77–88.

Delgado Rodrigues, J., 2001, Swelling behaviour of stones and its interest in conservation. An appraisal: Materiales de Construcción, v. 51, no. 263-264, p. 183–195.

Dunn, J.R., and Hudec, P.P., 1966, Water, clay, and rock soundness: The Ohio Journal of Science, v. 66, no. 2, p. 153–168.

Félix, C., 1994, Déformations de grès consécutives à leur consolidation avec un silicate d'éthyle (Deformation of sandstone following their consolidation with an ethyl silicate), *in* Proceedings of the 7th International IAEG Congress: Balkema, Rotterdam, p. 3543–3550.

Happel, J., Brenner, H., 1986, Low Reynolds number hydrodynamics: Dordrecht, Martinus Nijhoff, p. 389–404.

Jimenez Gonzalez, I., Higgins, M., and Scherer, G.W., 2002, Hygric Swelling of Portland Brownstone, *in* Vandiver, P.B., Goodway, M., and Mass, J.L., eds., Materials Issues in Art & Archaeology VI, MRS Symposium Proceedings 712: Warrendale, Pennsylvania, Materials Research Society, p. 21–27.

Katz, A.J., and Thompson, A.H., 1987, Prediction of rock electrical conductivity from mercury injection measurements: Journal of Geophysical Research, v. 2, no. B1, p. 599–607.

Kolsky, H., 1963, Stress waves in solids: Dover, Minneola, New York, p. 84–85.

MacEwan, D.M.C., and Wilson, M.J., 1980, Interlayer and intercalation complexes of clay minerals, *in* Brindley, G.W., and Brown, G., eds., Crystal structures of clay minerals and their X-ray identification: London, Mineralogical Society Monograph No. 5, p. 197–248.

Macey, H.H., 1942, Clay-water relationships and the internal mechanism of drying: Transactions of the British Ceramics Society, v. 41, no. 4, p. 73–121.

Madsen, F.T., and Müller-Vonmoos, M., 1989, The swelling behaviour of clays: Applied Clay Science, v. 4, p. 143–156, doi: 10.1016/0169-1317(89)90005-7.

McGreevy, J.P., and Smith, B.J., 1984, The possible role of clay minerals in salt weathering: CATENA, v. 11, p. 169–175, doi: 10.1016/0341-8162(84)90006-7.

Rodriguez-Navarro, C., Hansen, E., Sebastian, E., and Ginell, W.S., 1997, The role of clays in the decay of ancient Egyptian limestone sculptures: Journal of the American Institute of Conservation, v. 36, no. 2, p. 151–163.

Scherer, G.W., 1992, Bending of Gel Beams: method of characterizing mechanical properties and permeability: Journal of Non-Crystalline Solids, v. 142, no. 1-2, p. 18–35.

Scherer, G.W., 1999, Crystallization in pores: Cement and Concrete Research, v. 29, no. 8, p. 1347–1358, doi: 10.1016/S0008-8846(99)00002-2.

Scherer, G.W., 2000a, Measuring permeability of rigid materials by a beam-bending method: I. Theory: Journal of the American Ceramics Society, v. 83, no. 9, p. 2231–2239.

Scherer, G.W., 2000b, Thermal expansion kinetics: Method to measure permeability of cementitious materials: I, Theory: Journal of the American Ceramics Society, v. 83, no. 11, p. 2753–2761; Erratum, Journal of the American Ceramics Society, v. 87, no. 8 (2004) p. 1609–1610.

Snethlage, R., and Wendler, E., 1991, Surfactants and adherent silicon resins— New protective agents for natural stone, *in* Materials Research Society Symposium Proceedings: Pittsburgh, v. 185, p. 193–200.

Timoshenko, S., 1925, Analysis of bimetal thermostats: Journal of the Optical Society of America, v. 11, p. 233–255.

Tutuncu, A.N., Podio, A.L., Gregory, A.R., and Sharma, M.M., 1998, Nonlinear viscoelastic behavior of sedimentary rocks, Part I: Effect of frequency and strain amplitude: Geophysics, v. 63, no. 1, p. 184–194, doi: 10.1190/1.1444311.

Veniale, F., Setti, M., Rodriguez-Navarro, C., and Lodola, S., 2001, Procesos de alteración asociados al contenido de minerales arcillosos en materiales pétreos (Role of clay constituents in stone decay processes): Materiales de Construcción, v. 51, no. 263-264, p. 163–182.

Vichit-Vadakan, W., and Scherer, G.W., 2000, Measuring permeability of rigid materials by a beam-bending method: II. Porous Vycor: Journal of the American Ceramics Society, v. 83, no. 9, p. 2240–2245.

Vichit-Vadakan, W., and Scherer, G.W., 2002, Permeability of rigid materials by a beam-bending method: III. Cement Paste: Journal of the American Ceramics Society, v. 85, no. 6, p. 1537–1544.

MANUSCRIPT ACCEPTED BY THE SOCIETY 19 JANUARY 2005